EARTH SONG

A Prologue to History

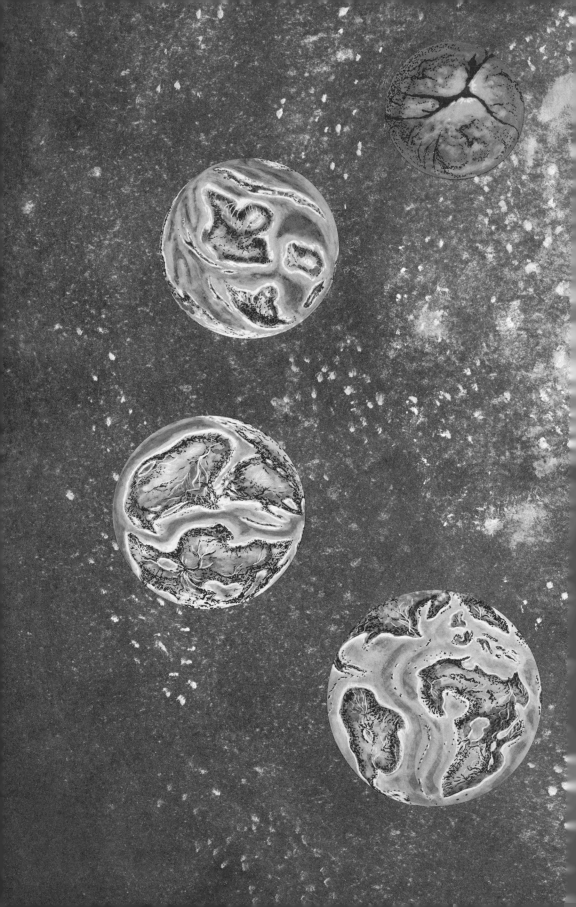

EARTH SONG

A Prologue to History

CHARLES L. CAMP

AMERICAN WEST
PUBLISHING COMPANY

PALO ALTO, CALIFORNIA

Library of Congress Card Number 71-128901

ISBN 0-910118-18-3

The original edition was copyrighted in 1952 by the
Regents of the University of California and published by
the University of California Press.

To

WILLIAM KING GREGORY

in deep appreciation

Preface

IN THIS OUTLINE of the history of land and life in the American West, details have been sacrificed. Some topics mentioned briefly are more fully discussed in the works listed at the end of this volume. To serve as a key to these references, parenthetic numbers have been inserted in the text.

The first edition of this book has been out of print for many years. Continuing inquiries and demand have made the publication of a new edition feasible. With the ever-increasing interest in the precarious position of man on this earth, it appears timely to bring into focus again the long, slow course of history of life in this changing world.

Suggestions and advice have been generously given by William H. Alexander, David R. Brower, Perry Bylerly, Jessie M. Camp, Harvey Fergusson, August Frugé, Edward W. Gifford, Robert F. Heizer, Dorothy H. Huggins, Robert M. Kleinpell, Alinda Macleod, Phil Ray, George R. Stewart, R. A. Stirton, Samuel P. Welles, Howel Williams, Jennie E. Woolley, Loren Eisley, and many others. Their kindness is gratefully acknowledged.

The plates were drawn by Margaret M. Colbert, the maps and numbered figures by Owen J. Poe, and the title page illustration by Gail Feazell.

Contents

CHAPTERS

Introduction 15

1. Earth Story 23
2. Rocks of Land and Sea . . 32
3. Life in Early Seas 38
4. Life Reaches Land 46
5. Age of Coal 52
6. Age of Reptiles 58
7. Age of Mammals 84
8. Coming of the Ice 102
9. Immigrants and
   First Families 106
10. Pleistocene Cemetery . . . 115
11. Rise of Man 123
12. Before the Mayflower . . . 133
13. Red Men of the Southwest . 137
14. Invasion 143
15. California on the
    Changing Earth 147
16. California's Changing Life . 156

Charts 163
Glossary 179
References 185
Index 189

INTERLUDES

Prelude 12
Record of the Rocks 20
Enigma 30
Tale of the Sea Rocks 36
Song of Salamanders 50
Song of Reptiles 56
Transformation 82
Deep Freeze 100
Pleistocene Parade 104
Black Death 112
Song of Man 120
Episode of Folsom Man . . . 126
Destiny 160

Plates

1. Early joint-necked fish, *Dunkleosteus*, from the Upper Devonian of Arizona 48

2. Lower Permian land vertebrates and landscape with the first pine trees *(Walchia)* 64

3. Upper Triassic vertebrates of Arizona, in forest of Auracarian pines partly destroyed by volcanic ash 71

4. Upper Jurassic life in Northern Utah 77

5. Upper Cretaceous reptiles of California, from the Moreno formation, west side of San Joaquin Valley . . 79

6. Early Tertiary mammals of California, from the Sespe beds, Ventura County 95

7. Landscape in early Oligocene at Titus Canyon near Death Valley 96

8. Giant bison *(Bison latifrons)*, Middle Pleistocene 108

9. Scene at Rancho La Brea showing birds and mammals whose remains have been found preserved in the asphalt 116

10. Folsom man observing bison *(bison taylori)* hunted by dire wolves *(canis dirus)* 129

11. A Chumash village on the Santa Barbara Channel . . . 140

12. Map of California showing localities mentioned in text 162

Prelude

California! Land thrice-born, cradled between the desert and the sea, hear the Earth Song! Song of the pulsing rocks — born of the heated earth, born from beneath the sea, and born in storms and floods upon the land.

Hear the Earth Song, O California! Song of the waves, swept from their ancient shores to give you birth. Song of the land delivered from the troughs of the sea by the labor of the trembling earth. Song of the mountains, rising in majesty above your sunlit strand, in green and tawny dress and ermine robes of snow.

Child of the rocks and waves, this song is sung for you.

Introduction

ARTH'S LIVING GARMENT, constantly renewed, changes in style through the ages. The tattered remnants of the past cover the earth with a frayed, patched pattern that recalls the styles of long ago. The rocks, like cemetery headstones, date these fashions that have come and gone through five hundred million years of earth history. Countless extinct forms of life, preserved as rock-bound fossils, bear witness to these changes in earth life. Fossils were embedded when the rocks were formed, layer upon layer, the younger lying above the older. The sequence of life is thus manifested in the sequence of the rocks.

Fossils testify to the works of creation. Their procession in the rocks represents the drama of history, unfolding the past ages of life, scene by scene: the long Age of Invertebrates and Simple Plants; the Age of Primitive Fishes, Amphibians, and Coal; the Age of Reptiles, Pines, and Cycads; and then the Age of Mammals with its backdrop of flowering plants.

The scene now being played is the first in a new act—the Age of Man. The play is not finished; the suspense is supreme, for we cannot look behind the dark curtain to see what new actors may be waiting to take the stage.

Each act of the life story with its inventions and achievements prepares the way for the next. New forms find new ways to live, adding their structures to the inventions of the past and using the old in new combinations to perform new functions. New forms of life expand over what has gone before, and build upon the past. The long history of early life merges into ours and helps us to comprehend our own.

Life, as we know it, is tied to the changing surface of the earth, on which it evidently arose, and where it continues to develop in profusion. The history of life in California, though only a small part of the greater picture, elucidates much that has happened elsewhere, for the procession of life, emerging from all lands and seas, has traversed and retraversed our land.

Most of the ancient life of California has vanished. Some forms now extinct here have survived elsewhere, and a few no longer living in other places have found refuge here—such relicts as the giant sequoia, the condor, and the night lizard of the desert.

California's life is changing, shuffling to keep in step with the ever-changing parade of living things, and adjusting itself to the varied and changing landscapes and climates. Her rejuvenated Sierra separates the eastern wastes from the central valley. Isolated coves lie tucked away in her Coast Ranges. The westerly winds break across these ranges and drop their moisture. A gamut of microclimates lies between the desert and the sea, on mountain slopes in zones marked by differences in temperature, and along the cool, north coast blanketed with summer fog.

Living things, adjusting their lives to these changing features, have been modified in subtle ways, in habits and in form. The redwood forests with their tan oaks, laurels, five-fingered ferns, and their coastal fringe of cypresses and bishop pines, extended down and up the coast when the fog belt presumably moved southward and then regressed to the north. Desert lizards, jumping rats, and tree yuccas invaded the interior valleys in hot, dry periods of the past and were left stranded there in small colonies along the western side of the San Joaquin. Mountain salamanders and lizards found their way southward along the ranges in cool periods and are now isolated on the southern highlands.

Gophers and fishes live at oases far out on the desert. In former wet periods their ranges were connected. Now, through separation and isolation, the populations have become diverse, and new species are evolving from the old. Song sparrows of the marshes have followed the advance and retreat of the marsh plants around the sinking shores of San Francisco Bay, until today they exist only in isolated colonies. And the birds have developed different songs and plumage in each of these small habitats.

Thus, the life of the land—from Indians to fishes, insects, and plants— reflects the diversity of California's topography and climate. This has attracted naturalists who have studied the influence of the isolation of small populations on the origin of species (21, 59).* Finally, students of heredity have investigated the fundamental nature of the changes involved in the formation of new features and have tried to understand how hereditary changes become perpetuated in living populations.

California, fresh from the hand of nature, and only recently encroached upon by civilized man, is a natural laboratory where active geological processes and changing life can be seen and studied to interpret the earth story. Fossils help to tell this story, and a few anecdotes will show how they have been discovered and what they signify.

At McArthur, Shasta County, California, a boy of thirteen was tending his father's sheep. He reached for a pebble to stimulate the lazy ones. To his surprise, that pebble proved to be a petrified bone, stuck fast in the gravel. He called his father, and between them they dug out a hollow horn core three feet long and twenty-two inches round the base. The question arose: What kind of animal had such a horn? Some called it a "cow critter"; but the horn was too big for a cow, or even for a long-horned steer. The University was notified, and an expert came to investigate. He unearthed the skull to which the horn belonged—a fossil skull of the extinct *Bison latifrons* (pl. 8), one of the largest and finest of its kind (64).

At the south base of Mount Diablo, a grading crew was widening the road at the entrance to the State Park. They uncovered an elephant tusk, ten feet long and a foot in diameter, buried under a layer of fossil soil eight feet down. Near this great seven-hundred-pound tusk lay the lower jaw and a bone from the three-foot hump of a big bison. And, strangely enough, the lower jaw of the Diablo bison fitted the skull from Shasta County two hundred miles away. The hump bone showed that this big bison had a great hump over his forequarters, like the smaller American bison living today. A gifted sculptor restored the huge bison head, which now rests in the Museum of Paleontology at the University of California, Berkeley.

The long-horned bison and the mammoth elephant inhabited California during the Ice Age when the mammals of the northern world reached their climax in size and diversity.

---

* Numbers in parentheses refer to the list of references at the end of the volume.

One hundred and seventy-seven years ago a Spanish botanist, José Longinos Martínez, traveled from Mexico City through Lower California to the little Spanish pueblo of Los Angeles (60). He found that the roofs of the adobe houses were covered with asphalt. Eight miles west of the village he visited some black pools where liquid asphalt bubbled from the ground. Dead birds and beasts lay in the pools, caught like flies on sticky paper. There were bones, too, of animals mired years before – animals long extinct and not previously known to have lived in California. There were elephants and mastodonts, giant saber-toothed cats and lion jaguars, ground sloths, tapirs, wolves, horses, camels, and bison (pl. 9). These tar-pit bison were not so huge as *latifrons* from Shasta, and they had longer horns than those that live today.

Bison of various kinds have lived in America only during the last three hundred thousand years or so. Their remains are numerous in old river channels on the plains of Nebraska; the big-horned ones are said to be the most ancient. In the course of time the horns have become shorter, and the modern, living bison has the shortest horns of all. Many kinds of bison were thus evolved.

Fossils of the animals that were buried first are in the deeper sediments and are the oldest. From the nature of organic remains and their positions in the rocks the ancient history of life can be deciphered and dated.

Allan Bennison, a high school boy, used to spend his weekends searching for fossils in the desolate hills along the west side of the San Joaquin Valley near Gustine. In the rocks he found sea shells and a few large bones, the remains of a fossil duck-billed dinosaur – the first dinosaur to be discovered in California. A few miles from the dinosaur, Allan also found the skull of a sea lizard, or mosasaur (12*b*). Soon afterward, geologists from Fresno State College discovered the skeleton of a seagoing reptilian monster, the remains of a swimming, fish-catching plesiosaur (67), a creature with paddles instead of legs, a neck fifteen feet long, and a small head (pl. 5). Because its teeth were specialized for catching fish and could not chew, the plesiosaur swallowed stones to help grind up the food in its belly. A hatful of smooth pebbles was found where its stomach had lain.

All these creatures—the dinosaur, the mosasaur, and the plesiosaur—have been extinct for about seventy million years. Nothing even faintly resembling them exists today. They lived long before the time of the tar pits and the deposition of the Shasta gravels where the big bison was found. And the fossils show that in the remote past, a hundred times as long ago as the Shasta bison, a very different world of life existed.

But even seventy million years is a short time in comparison with the whole extent of geologic history. Some fossils are ten times that old, and the most ancient rocks are at least thirty times as old as the remains of the California dinosaurs.

One hot summer day, two geologists and a student came down the dirt road through Cajon Pass from the Mojave Desert. They were thirsty and stopped under the sycamores by the stream where it runs through some ledges of ancient, upended rock squeezed between Old Baldy in the San Gabriel Range and San Bernardino Mountain. These ledges looked promising, and the visitors began cracking them with their hammers. Inside were some small black flecks. A magnifying glass showed these flecks to be the hard skins of creatures resembling pill bugs. They were trilobites, extinct relatives of the shrimps and crabs. These crawlers had lived nearly five hundred million years ago in seas that covered much of southern California at the time when living things first left their remains in abundance in the rocks. At this ancient time there were no fishes, reptiles, birds, or mammals, no land plants or insects. The sea contained the ancestors of all that lives today; for the life of today has been derived from those primitive forefathers—the invertebrates and aquatic plants of the Cambrian.

Through all the ancient ages, earth life streamed through California, sometimes leaving its remnants in the rocks. Some of the ages of earth history are represented meagerly in California, others are richly recorded. All the great life changes were reflected here, but great gaps interrupt the fossil story. One must go to many places, far and wide, where the fossils are better known, to fill the gaps of history. Excursions of this sort, like the search for ancestral graves, tend to broaden one's view and disclose much that has been concealed or destroyed in the story of our land.

Record of the Rocks

*O*ld Earth, worn by flood and fire, lifts her wrinkled face to the elements. Debris of mountains dead, decayed, and washed away encrusts her skin: tissue of battered rock, layered deep, damaged and crumpled, the older layers in fragments or destroyed; new, smoother mantles covering the old.

Internal tissues show in wounds — gashed canyons where deposits lie exposed — and primitive rocks congealed from the heated body of the earth. Movements under and within the crust have split and warped the skin, upraised the continents, depressed the ocean floors, and folded older layers into contorted forms. Ancient rocks lie fractured and obscure, hard to decipher and to reconstruct.

Rocks laid down by wind and water, layer on layer, entomb the creatures that lived long ago. Impressions, tracks, and skeletons appear when uplift exposes and erosion wears the walls of their stone sepulchers. These populations of the past, traced by their litter in the sediments, played their parts and passed away, to be succeeded by new multitudes.

Remains of life are scarce in older rocks. Few are the broken fragments and the spoils — algae, sponges, jellyfishes, the tracks and castings of worms — a mere suggestion of soft-bodied forms that left but little to be recognized. Buried in layers above, laid down in later times, are hosts of hard-shelled forms, ancestors of the clam and snail, forefathers of the crab and spiny urchin, the first fishes, and ancient waterweeds.

Life finally moved from sea and stream to land, changed in a thousand ways and gave rise to rooted plants with stems and leaves raised toward the sun; to air-breathing insects, spiders, and scorpions; to sluggish salamanders and their agile descendants, the reptiles.

Still later in the reptile age, scaly and leathery creatures spread across the land and grew monstrously in warm swamps. The deserts swarmed with *skittering and leaping forms. They swam in the seas and lakes. They flew in the air and stalked their prey in lush forests of giant ferns and monkey puzzle pines. The fleeter-footed reptiles gave rise to small hairy beasts and feathered birds.*

Bees came to pollinate the first blossoms, and ants to filch the fallen seeds. Families of termites clung together in the first well-ordered societies — workers, soldiers, queens — building permanent nests and laboring within the group; helpless to live again as individuals; bound into a new social order,

And now the mammals step upon the stage. For sixty million years they rule the land, and swim the sea, and fly the air together with the birds. Warm blood permits them to endure the northern cold. Intelligence dawns in the upright-walking, club-swinging man ape and culminates in man himself, set apart from other creatures by his curious, inventive ways; manufacturer of tools, user of fire, a scourge and terror to his kind and to the beasts which he exterminates, enslaves, or exalts to suit his whims and purposes. Intelligence dawns in the final scene of the last act, the Age of Mammals.

CHAPTER 1

Earth Story

EARTH, A PIN POINT IN THE UNIVERSE, contains the elemental things — atomic energy, all the common elements, and many of the rare ones. These elements occur throughout the universe, and their distribution seems to indicate that all the heavenly bodies were once connected in a common mass. The earth and the other planets were possibly thrown from the face of the sun, and the moon from the earth; but how this happened no one knows, and theories are conflicting. Perhaps they were drops pulled away by a celestial body that approached the sun and disturbed its surface. Perhaps the sun and the entire visible universe were exploded from a central mass and the earth was thus created along with all the rest.

Some astrophysicists and chemists favor the view that the solar system was condensed from a cold, cosmic cloud of gas and solid particles circulating around the sun (36). Such clouds may be seen in outer space today, obscuring parts of incandescent nebulae. The cloud-condensation theory seems to account more readily for the presence of water, the scarcity of rare gases, the uniformity of composition of the earth and the meteorites, and the presence of both metallic and oxidized iron in the earth. It also explains more plausibly the present distribution of the planets, with the smaller ones nearest and farthest from the sun. It assumes that the earth attained its present form and structure after condensation of the cloud, and that it became heated by radioactivity.

The spherical form of the rotating bodies of the solar system seems to show that they have gone through a gaseous or a liquid stage. The earth's interior seems to be arranged in concentric layers. The lighter materials at the surface, now solidified in a thin, solid crust, may have been formed from a siliceous scum that floated on a molten globe (16b). The lighter crust, some forty miles thick, resting upon a denser mass that flows like glacial ice under the great pressures exerted upon it, at times cracks violently, causing earthquakes. Molten matter, within a few miles of the surface, is often squeezed between layers of the cooler, crustal rocks and may reach

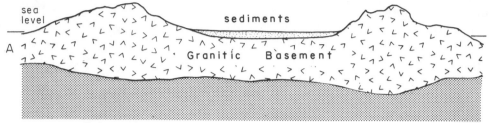

Thin sedimentary layers in sea trough

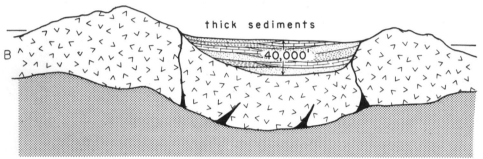

Sediments have accumulated to a thickness of 40,000';
basement has begun to sag and crack

Collapse of the crust

Fig. 1. Diagram of a subsiding geosyncline in which sediments are deposited until the earth crust bends and breaks. Magmas have entered fissures in basement and have formed volcanoes and lava flows in the newly upraised mountains.

the surface in volcanoes and through fissure vents. Molten rocks called magmas, as well as ash, cinders, and gases, rise to the surface, sometimes quietly, sometimes in violent explosions.

Explosions may cause earthquakes, but in California the principal cause of earthquakes is the rubbing together of colossal blocks of rock as they grind past one another in the fractured crust. The crust is but a fragile shell, full of deep cracks and breaks; in it the lines of stress and breakage often occur near the margins of the continents, as in California. Earthquake waves can be charted and timed as they pass through the earth and along its surface. The different waves are deflected and screened by the various densities of materials within the earth, and the behavior of these waves indicates something of the nature of the earth's mysterious interior (22).

The sea floor and the sea troughs and deltas, with their enormous loads of sediment constantly accumulating, tend to sink (fig. 1). The higher ground, from which sediment is being removed, becomes lighter and tends to rise; for the whole crust is supported by material that behaves like wax, and the lighter rocks float like icebergs in this heavy mass. The sea floors are also spreading apart from their mid-oceanic ridges. Europe and North America, and Africa and South America, were together about 275 million years ago, but have been drifting apart ever since. The great plates of the ocean floors drive against, and down under, the edges of the continental masses. This creates a weak zone of crushed and crumpled rock near the continental borders, the source of most of our earthquakes (1a, 46a). California, Alaska, Japan, the East Indies, the West Indies, the Andes, and eastern Africa are in such zones of movement and fracture. The great active faults of the Coast Ranges, the old volcanic fields stretching from the Cascades to the southwestern plateaus, Mount Shasta, the active volcano of Lassen Peak, and the recent cinder cones of the Sierra Nevada and the Mojave Desert have developed along such cracks.

Crustal movements have caused the sea to flood many times over the land in California. Much of what is now land has often lain beneath the sea. And some parts of the present continental shelf, now beneath the waves, were once dry land. Some of the rocks in the Berkeley Hills were transported from old granite mountains that stood where the Farallones still peer above the sea. The Coast Ranges and valley rocks contain sea shells and petroleum laid down at times when these regions were submerged.

Much of the surface of California today is a disordered mass of fragments of the past—detritus of mountains that no longer exist, carried away by rivers long extinct, into seas that have left only their traces on abandoned shores. Some of this debris, the most recent, still remains as unconsolidated sand and mud; some has become cemented by the percolation of lime and mineral waters. Thus, loose sand has formed sandstone, gravel has solidified into conglomerate, and mud has hardened into shale. These are the sedimentary deposits, which lie in thick strata in the bowl of the Great Valley and occur in broken, upended and folded masses in the Coast Ranges (fig. 19).

In the long course of earth history the older sediments often became deeply buried, and altered by heat and pressure, so that their original appearance was lost. Nearly all the Pre-Cambrian sediments were thus altered or metamorphosed, and a great mass of rocks of Mesozoic age in the Coast Ranges, the Franciscan series, is a mixture of old sediments, altered and interlayered with lavas and molten intrusions.

Of the three kinds of rock—the sedimentary, the metamorphic, and the igneous—the sedimentary (limestone, sandstone, and shale) is the most useful in preserving the history of life, for it carries by far the greater number of fossils. But the other rocks are valuable as a history of volcanism and mountain building.

The more recent sedimentary rocks rest mainly on old basements of complex crystalline material, largely granite. This material was formed thousands of feet below the present surface, by the solidification of molten masses which had formed under the cool crust or had been forced through its weaker parts. Wherever surface granites now appear, a mile or more of covering rock has been stripped away. The Sierra Nevada, the San Gabriel, San Bernardino, and San Jacinto mountains, and small parts of the central Coast Ranges to the east of Monterey are the granitic ruins or denuded cores of old uplifts that once bore a mantle of less resistant rock.

Much of California's terrain is rugged, folded and broken into hills and mountains of many kinds—upwarped ridges, fault blocks, erosional residues. Mountain remnants of great age lie in the hinterland, and toward the west younger ranges have been uplifted. Many of these have had a complex history. The Sierra Nevada, for example, originated about one hundred and thirty million years ago as a series of ridges fronting a gulf of the sea. A mass of molten granitic material was forced into or was formed within the old folded and broken sedimentary rocks, most of which have long since been worn away and carried off to the surrounding lowlands and the sea. This uplift ceased during the Cretaceous and early Tertiary periods, and the exposed granitic core became dissected into rounded hills traversed by river channels much shallower than the gorges of today.

Four or five million years ago a new outburst began with the rising of the eastern front along cracks which developed in the Owens Valley. This uplift of the eastern front gradually tilted the long Sierran block, raised the summit line about ten thousand feet, and rejuvenated the streams which have since cut deep, narrow canyons. Volcanic activity accompanied these uplifts. Craters and lava flows covered the slopes, blocked the stream courses, and buried the old, gold-bearing stream channels that had been left high and dry by the deepening of the gorges. Then came the glaciers of the Pleistocene, which polished the upper ridges, gouged and deepened the lake beds and canyons, and choked the lower stream beds with coarse debris.

Today, after the melting of the ice, the glacial lakes are being filled, forming meadows, the canyon cliffs are rapidly disintegrating and falling away, and forests are creeping toward the summit slopes. But the crests are probably higher now than they ever were, and small glaciers have again appeared in the shadows of the peaks.

While the highlands rose, detritus washed from them into adjacent seaways piled up to depths of more than thirty thousand feet. This heavy load caused the crust to collapse and heave into the complex ridges of the central Coast Ranges. Under this area lie old, contorted sediments deposited after the Sierra started to rise—sediments derived not only from the young Sierra but also from lands in the western sea.

The rise and fall of the coastal regions at different times and places enclosed embayments between the Sierra and the sea. These embayments continued to receive Sierran outwash during the Cretaceous and Tertiary periods, and now the sediments lie twenty-five thousand feet deep over the old granitic floor of the San Joaquin Valley, which has gradually sunk under the load but has not yet collapsed (fig. 19). The seacoast rose and fell, sea waters spread over the land and retreated, and lands arose in what are now the offshore platforms. Marine terraces or benches along the coast —at Pismo Beach, Santa Cruz, and elsewhere—represent the last phases of advance and retreat of the waters, possibly caused in part by the accumulation of glacial ice toward the poles, and in part by recent warping of the land (37).

The slow processes at work today in and on the earth crust and in the atmosphere are much the same as those that operated throughout the immense past. But the tempo of these processes has been stepped up in California, where the steep slopes of mountains are subject to more rapid weathering than are older, flatter landscapes.

History is a continuous stream. Cosmic history, earth history, fossil history, prehistory, and written history intermingle and overlap. They differ only in the nature of the documents by which they are studied. The complex pattern of the past merges into man's written records. The most accurate of all history is written in the rocks and lodged in the debris of the past. These fossil documents are so abundant that only a small number of them can ever be found and read. They extend from the very remote to the most recent times. Some are fragmentary, some are surprisingly complete. They are embedded in the land rocks and also in the sediment under the sea, which has been steadily accumulating and furnishes a record that is just beginning to be understood.

How slowly man has learned to look into the distant past. Only within the last hundred and seventy years has it been realized that the fossil record contains countless extinct forms of life. And only in the last one hundred years has it been shown that some of these extinct forms are the ancestors and builders of our present living world. Evolution has been demonstrated by the fossils as well as by the visible changes from generation to generation in living things. Yet this illuminating theory is frequently misunderstood, distrusted, and maligned — so much so that it cannot lawfully be taught or safely be discussed in some communities. Prejudice has seriously hindered the advance of knowledge throughout much of man's history.

Nevertheless, science sharpens her tools, devises new means of measuring, controlling, and perfecting her knowledge of nature. The fission of the atom is an awesome example of penetration into the vast unknown frontier of knowledge. Seventy years ago, soon after the discovery of the peculiar behavior of radium, it was thought feasible to measure the length of geologic time by the rate of decay of radioactive uranium contained in the rocks. This was an early application of the new knowledge of atomic decay. Previous estimates by various inexact means had given unsatisfactory and widely divergent datings of the time of formation of the crustal layers. Now, by weighing the residues of radioactive minerals in the earth crust, it has been found possible to date some of the rocks fairly accurately.

Active uranium, a natural clock, gives off radiant energy at a constant rate regardless of pressure or temperature. Radioactive uranium and thorium, when found in once-molten rocks in which they were originally enclosed and became crystallized, disintegrate into radium, residual lead, and helium. The lead accumulates in the disintegrating mass, and the helium is captured by certain common minerals; the two can thus be measured independently. The yearly rates of disintegration of uranium and thorium are known. The amounts of residual lead and helium therefore provide a means of determining the number of years the original mineral has been active (34, 65). An isotope of potassium, K^{40}, also has been found to be radioactive, and decays at a rate even more suitable for the measurement of geologic time (18a). Many dates from K^{40} have corroborated those based on uranium, and have so filled in the gaps in our time scale that it is reasonably accurate and internally consistent for the last 600 million years.

The uranium clock shows that the oldest rocks of the earth's crust were formed about four and one-half billion years ago. Manitoba granites have been given a date of two and a half billion years. The earth must be at least four and one-half billion years old, for these granites are intrusions in rocks containing pebbles of still older granites (chart 1). Since no fossils have been definitely recognized in the earliest rocks, there was probably no life on earth when these rocks were formed. But there are indications that simple organisms, the iron- and sulphur-reducing bacteria, were active during early stages of earth history. Deposits of bog iron in northern Michigan were evidently laid down by minute fossil bacteria which must be magnified many hundred times to be detected. These may be as old as 3.2 billion years.

Pre-Cambrian rocks (chart 1), which lack fossils, can usually be recognized by their position beneath layers which contain the earliest Cambrian life. Such rocks are often the residue of old mountain ranges. In this long period, mountains were many times raised and many times were worn away. The Pre-Cambrian record, distorted, fragmentary, and complex, cannot easily be traced.

Pre-Cambrian rocks appear in great masses beneath the lowest fossil-bearing sediments of the Grand Canyon. In California they are widely scattered across the eastern deserts; they are found in the Inyo Mountains, in the Tecopa region (40), and in the Panamint Range west of Death Valley, where Lower Cambrian fossils occur in the younger rocks lying upon the tilted and eroded surfaces of the Pre-Cambrian strata. The presence of old limestones in the Pre-Cambrian of this eastern district makes it seem likely that in this ancient time much of eastern and southern California lay beneath the sea. In the vast expanse of time before the Cambrian the land may have appeared again and again as the western parts of the continent rose and sank.

The complex Pre-Cambrian rocks are often placed in two periods, known as the Archean and the Algonkian. Rocks of the Archean period, which represents the first two thirds of geologic time, are chiefly volcanic and crystalline masses that formerly were molten. They lie beneath the remnants of the first known sediments. No recognizable fossils occur in them; but organic carbon, evidently the residue of early plants, has been identified by spectrographic tests of late Archean sediments. Rocks of the Algonkian, which represents the next five hundred million years of the record, are thick, contorted, and altered masses of sediments and old volcanic rocks. Fossils occur rarely, in the form of primitive plants, ancient animals such as glass sponges, jellyfishes, worm castings, and the microscopic, glassy skeletons of one-celled animals called Radiolaria. A long, still-unknown history of life must have preceded these forms.

Limy blobs resembling biscuits and honeycombs, found in the Algonkian of the Grand Canyon and elsewhere, seem to have been formed by ancient plants (algae) related to the modern pond scums and frog silk, and to the stoneworts—plants which form the tufa domes in brackish desert lakes such as Pyramid Lake and the extinct lakes on the Mojave Desert.

Enigma

Whence came the waters that lie in hollows on the earth, and whence the life within these waters? No water rested on the molten globe; no vapor of water encircled the primeval earth. The earth skin cooled to form bare, wrinkled crust, with many-colored rocks from pasty pulp and glassy mounds of slag. Did hydrous chemicals in union with the atmosphere produce these waters, or did the cooling earth permit the union of free oxygen and hydrogen?

By what uncanny alchemy did life arise, from pools of bubbling ooze— mixtures of hydrocarbons, tinctures of metals, salts, and strange acids, warmed by the heated earth, blown into spume by the winds, spattered, distilled, condensed—through billions of experiments in the broth of the first waters?

How long did Nature tend her sacred brews, patiently watching for the sign, the manifest immortal, seed of all to come, that corpuscle which feeds, respires, and grows, and duplicates itself?

Child of the hydrocarbon molecule, you first of Nature's host, destined to spread through sea and land and air in countless ways and forms, what was the meaning of your birth, and what will be your destiny?

CHAPTER 2

Rocks of Land and Sea

T HE ENERGY OF THE UNIVERSE shoots through space. Radiations, mainly from the sun, received upon this tiny earth, influence the cycles of the storms and, transformed, sustain life. The earth, seemingly floating free, is held and bathed in fields of energy. Electric, magnetic, gravitational, and radiant fields and forces cause the tides to ebb and flow, the waters to evaporate, the winds to blow, clouds to form, rains and snows to fall and melt, rivers to run, and rocks to disintegrate, dissolve, and wash away.

The earth itself, a pulsating body, sheathed in a vibrant atmosphere, covered with an ever-changing film of living matter, is subject to constant shifts and alterations in its heated interior and its cooler solid crust. The great plates of the spreading ocean floors push relentlessly into the continental margins, crumpling the rocks into parallel ridges and troughs. Loads of mud and sand, transported by winds and rivers from the weathered rocks, are dumped as soft, heavy masses into sea basins and across flood plains and deltas. Year after year, millions upon millions of tons pile up, until the thin crust below no longer can endure the growing weight. The crust bends and buckles and finally cracks, heaves, and collapses to form a mountainous mass of broken rock. Pressure forms mountains; these gradually wear away, and the rivers carry off the sediment; the whole process is then repeated at another place.

Much of the sedimentary blanket covering western North America accumulated in old sea troughs and embayments. Thick Pre-Cambrian sediments, volcanic masses, and outwash from mountains long since worn away, piled up time and time again in many an ancient basin. Paleozoic troughs, sinking to great depths under the sedimentary load, entered southern and northern California, Nevada, Utah, Colorado, Wyoming, Idaho, western Montana, western Canada, and southern Alaska. The troughs were more extensive than are the few surviving sediments.

Great seaways, in the Cambrian, crossed what is now southern and eastern California and continued through Nevada, Utah, and the Rocky Mountain region to the Arctic Sea and across central Alaska (fig. 2). This basin, called the *Cordilleran geosyncline,* continued to receive deposits during the whole early Paleozoic. It covered a wide area during the Mississippian. Lands uprose in western Nevada and in southeastern and southern California in the Pennsylvanian, and continental uplift climaxed in the Permian when most of California evidently became dry land.

Clear seas spread again over the northeast and the extreme south and across into Utah, in the Triassic (fig. 12); uplifts followed, in early Jurassic time, which confined the western seas to a long, narrow trough in the region of the present Cascades and the Sierra Nevada (fig. 15). Heavy deposition in this trough probably caused its collapse and contributed to the initial rise of the Sierra, in the Upper Jurassic. The sea then entered the newly restricted Franciscan gulf lying in the region of the present Coast Ranges, bounded by the new Sierra and by older granitic and volcanic ridges standing beyond the seaway, in the west (fig. 16).

Collapse of the Franciscan load of debris in turn formed ridges and embayments along the California coast, in the Cretaceous, and created a narrow sea with an eastern boundary along what is now the western margin of the San Joaquin Valley. This seaway was broader in the north, covering much of the Shasta-Klamath region. Lands still lay offshore to the west (fig. 20). Mild uplifts and depressions went on during the Tertiary. The ridges in the western sea diminished in height and extent. Then came the mighty upheavals of the Pleistocene, when the Coast Ranges were reborn and the present crest of the Sierra was reëlevated.

In the lost history of California, the wasting away and regrowth of the land occurred many, many times. The fractured crust received within and upon itself intrusions and extrusions from the molten subterranean materials. Over the surface, lavas flowed out and cooled; ash and pumice were scattered over the landscape. Subterranean molten masses infiltrated the solid crust and cooled to form granites and diorites streaked with the veins and ores now sought by the miners.

The greatest miner of all is the water which, falling from above, transports and abrades the surface debris. The soluble minerals and salts, as well as much solid material, are eventually carried off by the streams to the oceans, where they accumulate as in a reservoir and nourish the life of the sea. The living things of the sea utilize sea salts to build their skeletons. They change soluble compounds and gases into relatively insoluble substances such as chert and limestone. Corals and algae build vast reef-rock deposits. Microscopic shells form oozes that solidify into chert, chalk, and

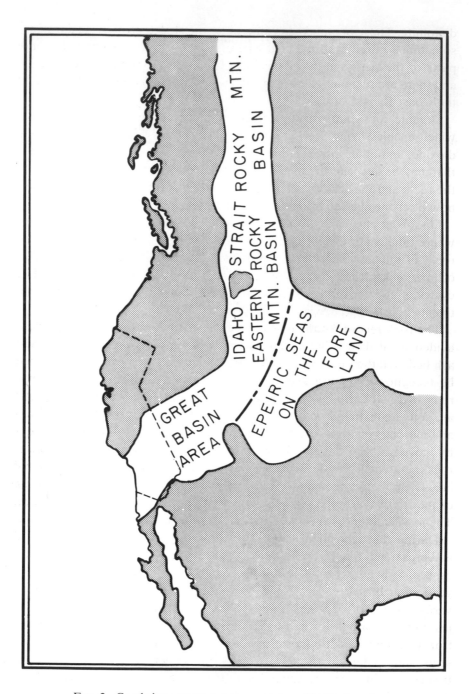

FIG. 2. Cambrian seaways across western North America. These troughs varied in extent during the Cambrian and the later Paleozoic. (From Deiss, 17.)

diatomite. These organic rocks formed by the skeletons of living things are sometimes raised onto the land when the old sea floor is uplifted. Large parts of the Alps and the Andes have been built from organic sediments, and there are remnant patches of such sediments in the Sierra Nevada and the Coast Ranges of California.

Organic rocks are usually formed in water by the energy of the sun, which enables living things to utilize soluble lime, silica, oxygen, carbon dioxide, and other inert materials, and to produce less soluble compounds that may be deposited when the creatures die and leave their hard parts on the reefs and the sea floor. Organic sediments formed by human activities now litter the surface of the land and the floor of the sea.

Dissolved chemicals from the sea water accumulate in shallow lagoons where tidal water evaporates. Beds of salt and gypsum are returned to the rocks of the land when these shore deposits are raised. Thousands of tons of lime and salt are carried into the ocean every year by the rivers. The cycles of erosion and deposition thus go on, and the solid crust of the continents constantly changes.

The rock records of the past can be read and translated, for they retain evidences of their origin and history. Old volcanoes, lavas, minerals, and ash beds tell their own stories. Ancient sand dunes solidified into rock can be recognized. Sun-cracked mud flats covered by the shifting sands of a river bed and carrying on their surfaces the tracks of prehistoric animals are sometimes found. Rocks changed by heat and pressure, such as slates and marble, rocks altered by subterranean decay, rocks broken and bent by movements of the earth's crust, rocks transported by wind and water, rocks scratched and gouged by glaciers, and rocks laid down in swamps and lakes help us to understand the fragmentary geologic record. The position of the rock layers, stacked one upon another, indicates their relative age. The fossil content of the rocks helps to solve their geologic age and history.

During the deposition of sediments, countless objects become buried and preserved. Shells, bones, wood — hard parts of animals and plants — are especially likely to become petrified and to be preserved for millions of years after burial. All sorts of remains and traces of life may be found in various kinds of deposits. Leaves, seeds, pollen grains; impressions of jellyfishes, worm trails and castings; carcasses of animals in ice and in frozen ground, in salt brine, and in peat; tracks in dune sand or in mud; insects and spiders in resin which turns to amber; soft animal and plant tissues in acid swamps; mummies, eggs and egg capsules; scales and ear stones of fishes; stomach stones of dinosaurs, plesiosaurs, crocodiles, and birds; animal fat in bogs; hair, dried ligaments, muscles, skin, and dung in caves — all these become parts of the fossil record, the instructive story of the history of life preserved in the rocks.

Tale of the Sea Rocks

*R*ocks, formed long ago beneath the waves, lie now all studded with the jewels of the deep—the pearly shells, the ruddy corals, the porcelain-coated teeth of sharks, the fragile sponges, and a trillion trillion diadems that once were skeletons of living pin points in the brine.

The scallop, the oyster, and the twisted whelk lie here, the chambered coils of nautilus and ammonite, spines of the prickly urchin, the pancake disc of the sand dollar, the calcite plates of the starfish, the crumbling feathery arms of the brittle star, the flower-formed heads and beaded stems of the stone lily, and the grandsires of the starfish tribe—the nutlike blastoids and the cystoids that live no longer in the waters. Here rest the broken body and the pincers of the crusty crab, the armored lobster and his squire the shrimp, and their forefather the trilobite. Here lies the lamp shell, decked in the translucent bonnet of a worm, style unchanging through the ages. A distant cousin of the brachiopod, the bryozoan, that grows like lacy moss, flourished in the ancient seas, and live today, reminders of the past.

Here lies the jellyfish, and here the pigmy polyps gather in colonies, the hydrozoans and anemones, blossoms of the sea—a feeble folk so weak in motion that they fasten down like plants upon the rocks and grow in cuplike skeletons of fiber and of lime. They cast their children forth as jelly blobs, to drift the sea, to found new settlements on far-flung reefs and distant shores where surging waves that batter mighty ships bathe and feed the fragile bodies of the coral and the sponge. Reefs of the past lie buried in the rocks where lay the warmer seas.

Here rest the swarms of midget shells, skeletons of lime and glass: the crystal jewel box of the diatom, precise in architecture and design; the perforated limy bulbs and whorls of the foram; the lacy globes and glassy spicules of the radiolarian. Here lie the stoneworts and the ancient seaweeds, crusted threads jointed and branching, bushy and bulbous forms that grew before the advent of the modern world and laid their debris down to form the limy rocks.

Here lie the ancestors of the fishes, and the later, stiff-backed, vertebrate aristocrats of land and sea, smooth and graceful in form and movement; swift to attack, to seize and bite and kill; perfected in their parts: nose, ears, eyes, jaws, teeth, and brain; muscles along a string of vertebrae and joined to rudder fins and swimming tail.

All these are laid away within earth's sepulcher, each in his place, so ordered and arranged that one can recognize the tribes and measure back through all the dynasties.

37

CHAPTER 3

Life in Early Seas

LIFE PERVADES THE SEA, the land, and the air, in variety beyond comprehension. Bacteria live in deep oil deposits and far down in rocks. Soil is formed by the action of small organisms. The atmosphere, in its lower thirty thousand feet, contains countless floating and flying forms. The sea teems with plants and animals.

Some of the viruses are scarcely larger than the largest molecules, and some are hardly smaller than the smallest bacteria. The viruses have some of the attributes of living things and have qualities of inanimate crystals as well. They may be intermediate between the living and the nonliving, but we know nothing concerning where, when, or how they arose, or whether they represent the forms from which life began. They exist today only as parasites in the cells of other organisms.

Living things exist by transformation of energy—the energy of chemical reactions activated by sunlight. The simplest living things have a semblance of immortality. They seem to exist continuously—to feed, digest, grow, regenerate destroyed or damaged parts, respire, secrete, excrete, reproduce by fission, and to evolve or change from generation to generation as they adjust themselves to changes in their surroundings. The substance of their bodies undergoes constant alteration and replacement. The form persists, and the substance continually passes through it like water in a whirlpool. The living cell is a complex of regimented molecules interacting chemically with one another.

The life substance, protoplasm, is a viscous fluid containing enzymes, salts, and acids of many kinds. Protoplasm cannot live in dry air unless it is enclosed in a protective coating; it must have arisen in a liquid medium. The simpler forms of life are still confined to a water-saturated environment during their active periods. Countless adjustments perfected the devices necessary for life on dry land and in the air.

The hereditary changes in living things, through countless generations, make their history interesting and useful. The series of such changes shown in fossils is so uniform throughout the world that one can usually come to understand the relationships of the rocks by comparing the fossils in widely separated regions. The numerous animals and plants of the early seas are particularly useful in the working out of these stratigraphic comparisons or correlations.

Layers of sediment laid down upon the earth and in the ancient seas are preserved without much distortion in the region of the Grand Canyon. There the lower layers were deposited before those that lie above (50). There, also, may be seen a succession of fossils in a visible sequence of rock strata. And comparisons can be made between this sequence of remains and similar sequences in other parts of the world.

Such comparisons show that certain kinds of organisms lived and ranged widely for relatively brief periods. Such fossils serve as useful markers in determining the age of beds not so thoroughly known. Thus, in California, where the early fossil record is poorly known, a few recognizable markers give important clues to the succession of faunas and the ages of the rocks. Little could be established in California were it not possible to make comparisons with better-known records elsewhere.

The story of life on earth as recorded by abundant fossils begins abruptly in the Cambrian some five hundred million years ago, when living creatures with hard shells and hard, horny skins came on the scene suddenly and in great abundance. Pre-Cambrian rocks contain few fossils; the sudden appearance of the hard-shelled invertebrates at the beginning of Cambrian time is a puzzle. Perhaps the ancient waters of the sea did not provide chemicals in proper balance for the formation of hard skeletons. Perhaps predatory invertebrates did not develop until late, and the rise of shelled and armored forms was a response to the need for protection against the new predators. Life must have existed previously in rich abundance and variety, but naturally only the organisms with hard skeletons or shells are likely to be found in the rocks. Only a few early, soft-bodied fossil forms are known, preserved of course in exceptionally fine-grained sediments. The Cambrian is the first period of the Paleozoic era; it embraced an estimated 120 million years.

The life of this ancient time has left its mark in many parts of the world; there are similar features everywhere. Fossils of simple, floating, and lime-

FIG. 3. Cladophoran algae.
A. *Cladophora gracilis*, living form.
B. *Marpolia spissa* from the Middle
Cambrian Burgess Shale, Mount Field, B. C.
These filamentous blue-green algae
have changed little through the ages.
(After Walcott, Smithson. Misc. Coll.,
67, pl. 51, fig. 1, and pl. 52, fig. 1.)

A

B

secreting plants distantly related to the seaweeds (fig. 3) have been found, but there are no records of land plants in this period. Single-celled microscopic animals (Protozoa), which secreted hard shells or glassy, jewel-like skeletons (fig. 4), and tube worms (58) lived in these ancient seas. Delicate sponges (fig. 5) supported by granules of lime and vaselike networks of translucent spicules, jellyfishes, graptolites, solitary cup-shaped corals (fig. 9), the nutlike ancestors of the crinoids or stone lilies, small lamp shells (brachiopods), archaic snails—all these made their debut in the Cambrian.

The most highly developed creatures were the now-extinct trilobites, distant relatives of the shrimps and sowbugs of today; like the sowbugs they had a horny, jointed shell, jointed legs, good sense organs, and coordinated muscles which permitted them to roll themselves into a "pill." The trilobites crawled over the sea bottom and probably fed, as crabs do, on refuse. Their remains are abundant, all the more so because the trilobites,

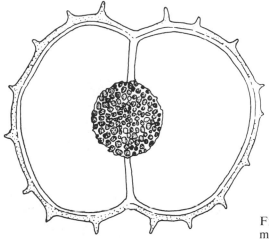

FIG. 4. Glassy skeleton of minute radiolarian protozoan, *Spongosaturnalis spiniferus*, from Upper Cretaceous limestone near Tesla, Calif. (After Campbell and Clark, Geol. Soc. Amer. Special Papers, 57, pl. 2, fig. 4.)

like the crabs, shed their hard skins during growth. Each left a potential fossil every time it shed its horny shell.

One of the short, broad-bodied trilobites *(Olenellus)* carried sharp, recurved spines on the sides of the headshield and body (fig. 6). *Olenellus,* a useful fossil marker of the Lower Cambrian (Muav) in the Grand Canyon, is one of the few fossils known from the Cambrian (Inyo, Marble, and Providence mountains) of California; fossils of this period were laid down in the sea along ancient troughs (17) extending from southern California across Nevada and Utah, and far north into Canada (fig. 2).

The Ordovician was the second period of the Paleozoic; its duration is roughly estimated at forty-five million years. It was the period of greatest continental submergence. Sea troughs, as in the Cambrian, lay in the Cordilleran geosyncline (fig. 7). Beds of this age, laid down in ancient seaways, have been recognized in California only in the Inyo Mountains and farther east in the Panamint and Telescope ranges and in the Nopah hills near Shoshone.

The life of the Ordovician was more varied and more advanced than that of the Cambrian; it included several new groups of animals, as well as all the old ones. Glass sponges were numerous. Lime-forming polyps (hydrozoans) and colonial corals appeared. The graptolites — seaweed-like animals that lived in clusters of feathery sprays with floating devices like water wings — have left fossils that are useful as markers, remarkable in abundance

Section

FIG. 5. Skeleton of early sponge, *Ajacicyathus nevadensis*, from Lower Cambrian beds, Silver Peak, Nev., and White Mts., Inyo Co., Calif. The delicate skeleton is formed of calcareous granules; other sponges at this time were formed of glass spicules. (After Okulitch, Geol. Soc. Amer. Special Papers, 48, fig. 18.)

and variety (fig. 7). The first stone lilies are found in the Ordovician record, along with their ancient relatives the blastoids—descendants of the nut-like cystoids of the Cambrian. Here, too, are the first sea urchins. Masses of fossil worm tubes lie entwined in coils, and, plastered over rocks and shells, is the ornamental tracery of the moss animals, the bryozoans. Hinged brachiopods took the lead over their old Cambrian relatives, the unhinged forms, in the Ordovician; spiriferous brachiopods and a few small, simple clams emerged. There was a great increase in the snail populations. Nautiloids, related to the relict chambered nautilus of today, arose. Some of these were coiled, others *(Orthoceras)* had straight shells which reached a length of fifteen feet. The trilobites became more varied and larger, some more than two feet long (45*b*).

Fierce predatory animals of the Ordovician were the giant scorpion-like eurypterids, which were barely represented in the Cambrian. They doubtless preyed upon the ancestors of the vertebrates—the plated ostracoderms. Ostracoderms, earliest of the backboned animals, antedated the fishes; the

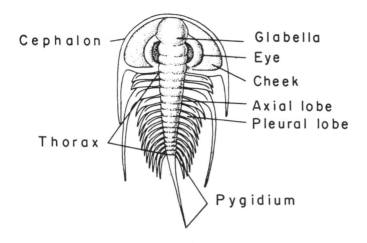

Cephalon

Thorax

Glabella
Eye
Cheek
Axial lobe
Pleural lobe

Pygidium

FIG. 6. Trilobite crustacean, *Olenellus gilberti,* from Pioche, Nev., Middle Cambrian. Besides the structures shown here, there were simple antennae, a pair of jointed, crablike walking legs on each body segment, and feathery legs (leg-gills) used for swimming and breathing. (After Walcott, Bull. U.S. Geol. Surv., No. 30, pl. 21, fig. 1.)

oldest ones are found in the Upper Ordovician Harding Sandstone of Colorado. They were armored and sluggish, skatelike and fishlike in form, fundamentally similar to the modern lampreys and hagfishes, which differ widely from the true fishes in lacking jaws (51).

Rocks formed in the Silurian period in California are poorly known. In the Taylorsville district of the northern Sierra, small bodies of limestone contain corals like those in the Niagara gorge in New York state, of acknowledged Middle Silurian age. Dolomitic limestones east of Towne Pass in the Panamint Range of Inyo County contain the biscuit coral *(Stromatopora)* of Silurian age (31). A more abundant Silurian fauna has recently been discovered in Shasta County but has not yet been studied. In that period, seas evidently still occupied the inland embayments which had existed since Cambrian time, but their boundaries had changed.

Much of the Silurian life resembled that of the Ordovician. One of the novelty acts was the coming of the scorpions. All scorpions are now land animals, but these ancient ones may have been aquatic. A four-sided pyra-

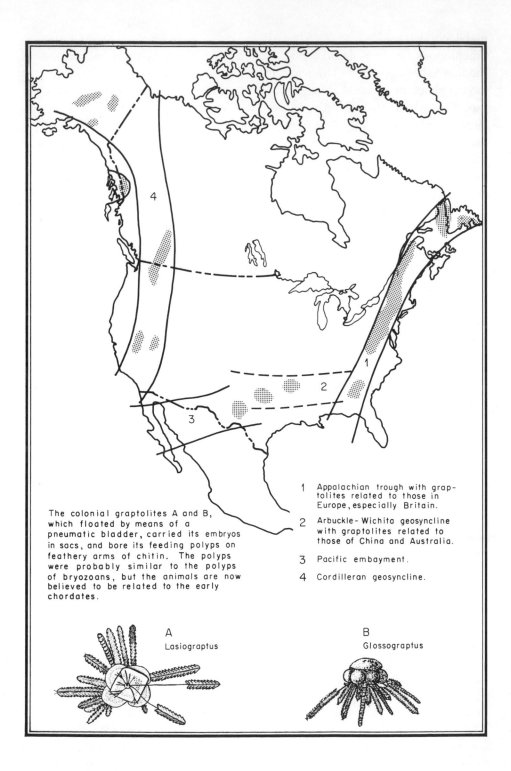

The colonial graptolites A and B,
which floated by means of a
pneumatic bladder, carried its embryos
in sacs, and bore its feeding polyps on
feathery arms of chitin. The polyps
were probably similar to the polyps
of bryozoans, but the animals are now
believed to be related to the early
chordates.

1 Appalachian trough with graptolites related to those in
  Europe, especially Britain.

2 Arbuckle-Wichita geosyncline
  with graptolites related to
  those of China and Australia.

3 Pacific embayment.

4 Cordilleran geosyncline.

A
Lasiograptus

B
Glossograptus

midal coral *(Goniophyllum)* is characteristic of the Silurian. Spiriferous brachiopods became abundant, and the first of the placoderm and acanthodian fishes appeared. The placoderm fishes differed from later forms in having plated bodies and jointed heads. The acanthodians had weak rudder fins. This was the time when fishes took on rapacious habits, feeding on large live prey and using their newly acquired jaws for this purpose. Before the advent of jaws, the ostracoderms *(Cephalaspis)* had small, soft mouths, useful only to sift out microscopic organisms. And this was the time when the fishes innovated bony jaws, rudder fins, and teeth, which they handed down to their numerous vertebrate descendants.

FIG. 7. Ordovician and Silurian sea troughs in North America, showing distribution of graptolite beds. (After Ruedemann, Geol. Soc. Amer. Mem. 19, fig. 1, *d,* and pl. 78, fig. 19; and pl. 82, fig. 19.)

CHAPTER 4

Life Reaches Land

Rocks of Devonian age are reported in Trinity County, in the Klamath Mountains (Redding district), in the northern Sierra Nevada near Taylorsville, in the Inyo Mountains, in the Panamint and Nopah ranges, and through Nevada and Idaho. The lower sandstones and conglomerates in the eastern desert areas may be of fresh-water origin. The rest are probably marine (42).

Near Payson, Jerome, and Flagstaff, in central Arizona, thin beds of pink limestone and heavy conglomerates contain fossils of large, predatory arthrodiran fishes *(Dinichthys)* like those found in the Cleveland shale of Ohio (pl. 1). Above these fish beds, which lack marine fossils, are extensive limestones with fossils of sea-living invertebrates.

The Devonian, which began more than three hundred million years ago, marks the beginning of the record of land plants. Plants previously were simple aquatic forms without leaves, stems, or roots. In this period, plants for the first time achieved vascular stems, leaves, and simple roots — structures developed for life on land and for drawing moisture from the soil. In changing from aquatic to terrestrial habitats, the plants developed supporting tissues and a leathery cuticle for protection against desiccation.

The earliest land plants (psilophytes) are intermediate between the mosses and the ferns. Competition for sunlight urged these primitive land plants to grow higher and higher, until they eventually formed trees and forests. Smaller ferns and horsetails were content with the subdued light on the dark forest floors and formed the undercover as they do today (3). Ferns of many types, even giant tree ferns, grew in the first forests. The club mosses, small plants today, developed into big, scale-scarred, leathery-barked trees — trees with soft spongy tissues in their trunks, yet extending to a good height and thickness and growing in dense forests (33).

Among the Devonian invertebrates, the sponges began to secrete more solid, limy skeletons, although many of the old, glassy types remained. Graptolites became almost extinct; the hydrozoans (stromatoporoids) reached their climax. Corals became more abundant, and complex reef-building forms occurred in profusion. Crinoids seem to have developed into free-swimming and crawling, as well as sedentary, types. Blastoids practically disappeared after a late burst of development. Only a few cystoids, the earliest of the echinoderms, remained. Starfishes arose, and sea urchins became abundant. Brachiopods reached the peak of their development. The clams (pelecypods) became diverse, and the nautiloids (cephalopods) became advanced in variety of structure, color, and ornamentation of the shell. Primitive ammonoids appeared (45a).

The Devonian witnessed the extinction of the ancient ostracoderms and the beginnings of true fishes, sharks, lungfishes, crossopts (fringe fins), and porcelain-scaled ganoids, preserved in the Old Red Sandstone of Scotland made famous in the writings of Hugh Miller. The successful advance of the crossopts toward land life is important. The Old Red of Scotland and Norway is thought to have been deposited in a widespread series of freshwater lakes thickly populated with ancient fishes, among which were many that could rise to the surface of stagnant waters and gulp air into lungs as the lungfishes do today. Some of the modern lung-fishes have developed extraordinary devices to prevent death during times of drought when the shallow lakes dry up. They live now only in the lakes and streams of Africa, South America, and Australia, and they are adapted to stagnant waters where other fishes die for lack of oxygen. The African *Protopterus* encases itself in a cocoon in the mud to aestivate during the dry season, and breathes into its lungs through a "straw" secreted by glands in the mouth.

Continental uplift in the late Devonian created arid climates. Streams and lakes became intermittent; and some of the air-breathing fishes lost the use of their gills, breathed through nostrils into their lungs, and finally used their fins as feet. In the shallow lake and stream beds they could wriggle over the mud to seek low spots where water still remained (51). The late Devonian rocks of east Greenland contain fossils of a peculiar salamander-like amphibian which retained the skull form of a crossopt fish, except that it had no gills. This was the first of the land-living group of animals to which belong the salamanders, frogs, and toads. And it was in this period

PLATE 1. Early joint-necked fish (*Dunkleosteus*) from the Upper Devonian of Arizona. This, the largest of the early plated fishes, belongs to a group which was the first to develop jaws, teeth, and predatory habits. It reached a length of about thirty feet.

that vertebrates first developed the walking feet that enabled them to move about on land.

During the evolution of structures suitable for land life, many millions of creatures must have perished in attempts to adjust themselves to the new habitat. The desperate struggles of present-day salamander larvae and fishes to escape from the mud of a drying pond are reminders of these earlier struggles for existence.

Earth is small for the life that swarms upon it, seeking food, wrestling with the elements, moving from place to place, rapidly reproducing. Crowded life struggles to survive, to find new ways to live, new things to feed upon, new physical parts, and new ways to use the old parts. Changing seasons and climates press and exterminate. The land changes; floods and drought, heat and frost bring starvation and death. Life has consequently crowded into strange places and has become curiously modified. Life from the sea entered the streams and lakes long ago, and these fresh waters served as a gateway to the land. For scarcely anything came directly from the sea to land without a course of training in the fresh waters.

The crab and scorpion, snail and insect tribes, found ways to change their breathing gills to live on land. The active fishes perished in the vanished ponds or smothered in stagnant water holes along once-running streams. But fishes that gulped air into their lungs survived this trial and could wriggle out of drying spots toward larger, deeper pools, there to hatch broods of better wrigglers — which eventually would walk and use their fins as clumsy feet. Hence, in Devonian time, arose the ancestors of all the footed hosts that range the land: frogs, lizards, beasts, and finally birds.

The pioneering plants, first to win their way to land, developed into forest trees of ancient form. Footed amphibian creatures then crept forth, hugging the moister ground and dallying long before they left the swamps to venture further toward the land.

Song of Salamanders

Swamp spawn of the sluggish fishes, flat-headed monsters skidding through the reeds like sleds; bodies broad, short-necked, propelled by feeble legs; slippery scales on leathery hides; rough, pitted bones to bind hides fast to head and chest—these are the water beasts that sprawl in shadowy pools, their pin-toothed mouths agape to snap the hapless fish. These are the first of the land travelers—swamp salamanders, with ears to catch the mating calls resounding through the forests of scale trees.

Egg-bloated bellies slither toward the breeding pools where bobbling bulls gather for courtship frolic; urging, nudging, pressing, embracing, arousing their torpid mates to shed the eggs amid the scattered sperm. Egg clusters, glairy capsules, catch the sunlight, hatch in the sun heat. Wiggling tadpoles break the jelly pouches to emerge and swim, to breathe through gills, as fishes do.

Horny-beaked, they nibble silken algae in the pool. Potbellied, they lie lazy in the sun, to grow and, as they grow, to change. Gills shrivel, nostrils perforate the skull, permitting air to pass into the lungs. Tails waste away, legs bud and sprout, bone teeth replace the horny dental plates. Coiled entrails straighten to digest fleshy food.

Of teeming thousands, few survive. These plod across the land from stream and lake and swamp, an awkward lot—moist, shiny, rubber-skinned; some with round, pancake heads, some broader still, and some with tapering, quick-snapping snouts. All meet the challenge of the land with crude experiments—trussed columns to support the body weight, and union of the pelvis with the vertebrae. The most advanced produced the reptile tribes—breeding on land, quicker in movement, with lighter heads on longer necks to seize more active prey, and dry-skinned to survive the droughts.

CHAPTER 5

Age of Coal

M OST OF NORTHERN CALIFORNIA during the Carboniferous, the Age of Coal and Amphibians, was covered by the sea. There is no record here of land life in this period, and no suggestion of the lush coal forests and swamps, so common elsewhere in the Northern Hemisphere. A small land area existed for a time in the Klamath region where volcanic ash and tuff (Baird formation) lie beneath the McCloud limestones of Permian age (48).

Conglomerates and sandstones in the northern Sierra and in the Argus Range and Inyo Mountains may have been washed from ancient highlands in central Nevada. Limestones at least forty-five hundred feet thick in the northern Argus Range contain corals and other marine fossils identified as Mississippian (Lower Carboniferous). Equivalent beds occur in many of the Mojave Desert ranges and in southern Nevada, and possibly also in the San Bernardino Mountains (26).

Pennsylvanian fossils, through eight thousand feet of strata, overlie Mississippian limestone in the Darwin Hills and the Argus Range. The upper part of the sequence probably extends into the Permian, for it contains the foraminifer *Pseudoschwagerina* (fig. 8).

The ancient Carboniferous seas that spread across California were inhabited by creatures of warm, clear waters—crinoids, corals, brachiopods, and foraminifers—just as in many other parts of the northern world.

Fig. 8. Shell of fusulinid foraminifer, *Pseudoschwagerina arta,* from Bird Spring formation, Lower Permian, Providence Mts., Calif. Forams such as these are widely used in correlating the marine Permian rocks. (After Thompson and Hazzard, Geol. Soc. Amer. Mem. 17, pl. 18, fig. 2; sagittal section.)

The Southern Hemisphere, at the close of the period, was invaded by successive sheets of ice.

Land life elsewhere flourished during the Carboniferous. There were widespread, cool, uniform climates and low-lying swamps in which grew the mighty coal forests of Europe, eastern North America, and China. Here the ferns and seed-fern trees reached their climax in form and variety. The club-moss trees (lycopods) and the horsetail trees (calamites) made up a large part of the coal forests. Their roots were shallow, their weak, pithy trunks were enclosed only in a thin, leathery bark; they could have endured neither battering of heavy winds nor freezing cold. The climates then, over a wide belt in the Northern Hemisphere, must have been mild and equable. And they were probably cool, with fairly constant rainfall throughout the year: no seasonal rings were formed in the woody trees.

The coal forests were new in the world. They had started with the early Devonian, grotesquely branched, seed-fern trees and the tree ferns of the late Devonian, and they reached their climax in the Carboniferous, when a variety of larger trees appeared—the lycopod scale trees, horsetail trees, and cordaites with strap-shaped leaves. The last had thicker woody tissues in a ring around the central pith—tissues that later grew to form the solid trunks of the pines of the Permian and Mesozoic.

This lush vegetation used the sun's energy to combine the carbon dioxide

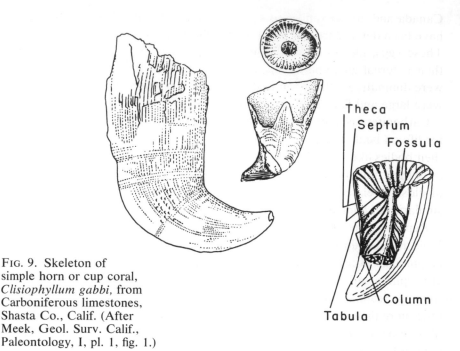

Theca
Septum
Fossula
Column
Tabula

FIG. 9. Skeleton of simple horn or cup coral, *Clisiophyllum gabbi,* from Carboniferous limestones, Shasta Co., Calif. (After Meek, Geol. Surv. Calif., Paleontology, I, pl. 1, fig. 1.)

gas of the atmosphere with water in the leaves of the plants. Each green leaf intercepting the light was a chemical laboratory in which the plant manufactured its own food. Some of the foods were used for the growth of the plant and for the manufacture of the woody cell walls (cellulose) of the plant tissues. Plants contain weak acids, such as tannic and acetic acid. When woody detritus such as leaves, stems, and trunks fell into the still waters of the Carboniferous swamps, the waters became so charged with these acids that bacterial decay was hindered. The dead plant tissues piled up in those ancient swamps, and were sometimes covered and compressed by mud and limestone. Thus they became altered and preserved as coal. Coal, like petroleum, contains stored-up energy from the sun, transformed by plants.

The West has no coal of Carboniferous age. Most of the coal of Utah and New Mexico was laid down in the Cretaceous, and the small coal beds near Mount Diablo, east of San Francisco, are of still later, Eocene age. Petroleum and gas deposits, which are elsewhere abundant in the Carboniferous, were formed in California much later (49).

The great fresh-water swamps, lakes, and forests of the Carboniferous sheltered two-foot dragonflies, four-inch cockroaches, and amphibians which differed from the salamanders of today in having skulls solidly roofed with bone. They arose in the late Devonian of Greenland and

Canada and are so similar to some of the Devonian fishes that they must have been derived from them. Like the fishes, they laid their eggs in water. These eggs, like those of the lungfishes, hatched as tadpoles, which lost their external gills and changed their structure as they grew. Some forms were diminutive, others had bodies up to fifteen feet in length. Their heads were large and flat, with huge mouths that could have been used as traps.

Carboniferous amphibians were a water-loving lot, poorly developed for life on land. Their stubby limbs and weak feet barely served to propel their bodies like sleds through the swamps. Their skins had to remain moist, as in their modern relatives the salamanders and the frogs. The occurrence of closely related forms in both North America and Europe indicates a similarity of climate and close continental connections.

Reptiles were the first vertebrates to succeed well in life on dry land. Land reptiles do not have to return to the water to breed, as most of the amphibians do. Their dry, scaly, and leathery skins can withstand a dry atmosphere and desert conditions. Reptiles are able to get about on land by running, leaping, and flying; whereas the ancient amphibians could do little more than crawl and swim and lie in wait for their prey. The first reptiles, in the Pennsylvanian, may have lived more happily in the dry uplands than in the coal swamps.

Song of Reptiles

Ancient brood: first vertebrates to run and leap and fly, peopling the earth with varied forms, leaving the waters to stride boldly forth on land; and first to breed on land!

Sluggish Devonian fishes, drought-stricken, left the drying ponds, on stumpy fins, to push their bodies through the muck. Creeping salamanders changed fins to legs and feet, and trudged on feeble limbs through the damp forests. In this Age of Coal, reptiles emerged, offspring of salamander tribes, to win the land and find new ways to move and feed and propagate.

Conquest of land was difficult and slow, a long and patient struggle by the scaly ones. Each new invention of swift foot or crushing jaw, brain part to serve a quickened sense of sight or sound, muscles and joints to give increased mobility, was won at bitter cost. Battalions, inept and crude, were swept away by companies of abler forms bearing new weapons, some new device or skill that gave supremacy.

Reptiles learned to crunch tender plants, to dig the juicy rhizomes of the ferns, to dredge the ponds for soft-stemmed water weeds, to tear the husks from stems and pods, and to climb trees for buds. Others developed dagger teeth, and active limbs with which to spring at sluggish herbivores, to snatch the giant dragonfly and wily cockroach skittering through the leaves. Old sluggish herbivores, cotylosaurs, declined, leaving the plodding turtles as the sole survivors of their clan. The active carnivores produced the mammal hordes.

Some reptiles learned to plunge and swim, to crush the feeble mollusk in his shell, to grind the crab to bits within his crackly skin, to foil the fishes and the darting squid. These ventured to the sea, and from them rose the varied hosts, whalelike and porpoise forms—ichthyosaurs, plesiosaurs, mosasaurs, and seagoing crocodiles.

Ancestral two-legged dinosaurs, handy paws held ready, combed the beaches, patrolled the mud flats, scampered over the dunes, sprang on their prey and flounced aloft as they ran. From them evolved the birds and monstrous swamp giants: brontosaurs, camptosaurs, titanosaurs. From them descended the outlandish armored forms of the Cretaceous: shapes of elephantine tortoises and rhinos; plated, horned, squat, saddle-necked, pipe-nosed, hoofed, clawed, biped, and duck-billed types.

Some dinosaur-like agile ones parachuted from trees, like flying squirrels. They led the way toward fluttering, macabre forms, the pterosaurs; of these, the giant argonaut, Pteranodon, skimmed the Cretaceous seas on steady, soaring wings.

The mesosaurs grew teeth like bristles on their slender jaws, to sift and strain tiny plankton from the brackish seas. The lizards learned to swim in sand, to burrow and traverse their burrows fore and aft with equal speed. Lizards gave rise to snakes that slink, glide, coil, and strike with deadly fangs—snakes that engulf their prey by monstrous stretching of their jaws.

All these and many more lived, and gave way to furry beasts, warm-blooded, insulated from the cold, able to gnaw and masticate, to digest food more effectively, to range the winter snow, to move with more endurance and agility.

CHAPTER 6

Age of Reptiles

THE PERMIAN WAS USHERED IN by the rising Appalachians and by a simultaneous uplift of highlands across Europe and Asia. Rivers flowing from these heights filled the interior valleys and the coastal embayments with sediment and spread over broad flood plains along the Gulf of Mexico and the old Pacific and Atlantic seaboards.

Northern California lay beneath a temperate sea where the McCloud limestones were deposited. Permian strata in the southeast are remarkably thin in the few places they occur—the Owenyo limestones in the Inyo Mountains, the Darwin Hills, the Argus Range, and in the Providence and Marble mountains (25). A seaway probably extended across Arizona into what is now the region of the Grand Canyon where the Kaibab and Toroweap limestones, now at the Canyon Rim, were laid down. These limestones thin out to the east and west. Below them are thick, Lower Permian continental sediments which grade into marine beds in western Arizona and Nevada. The shifting shore line ran north-south through Arizona and Utah.

The conifers—ancestors of the modern pines, spruces, redwoods, yews, and cypresses—first flourished in the Permian. These trees, with their compressed, needle-like leaves, were fitted to withstand drought. Their stout, woody trunks were strengthened for growth on windy uplands where the ancient coal forests could not exist. The strap-leaved cordaites and other, more progressive, coal-forest trees survived into the Permian, along with the calamites and a few relict lycopods. After the great glaciation in the Southern Hemisphere, a group of broad-leaved plants (glossopterids) arose and gradually spread to the north, even to Greenland at the end of the Triassic. In Arizona, other types of broad-leaved marsh plants, related

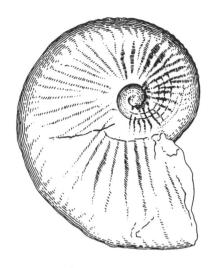

FIG. 10. Ceratite ammonoid shell, *Trachyceras californicum*, from the Upper Triassic Hosselkus limestone, Brock Mountain, Shasta Co., Calif. Wavy line shows one of complicated sutures which join shell chambers. (Described and figured by J. P. Smith, U.S. Geol. Surv. Prof. Paper 141, pl. 1.)

to those in the coal beds of China, grew in the Permian.

California sea life, as early as Devonian time, was more closely related to that of eastern Asia than to that of middle and eastern North America. The marine Permian invertebrates were more like those that inhabited the seaways over what is now Siberia.

Highly developed, internally coiled foraminifers (fusulinids), abundant in the Permian, are useful now in identifying levels from rock cores drilled in the oil wells. There were but few corals in the Permian, which possibly indicates a general cooling of the sea. The blastoid echinoderms disappeared in this period. Spines and plates of sea cucumbers (holothurians) occurred for the first time. There were large numbers of shellfish: clams, sea snails and brachiopods. Ceratite ammonoids (fig. 10) by this time had diverged from their nautiloid ancestors (fig. 11). Trilobites became extinct, also the ancient grandfathers of the cockroaches (Paleodictyoptera), the first winged insects of the early Pennsylvanian (10).

The early Permian land-laid rocks of Texas and New Mexico alternate vertically with marine limestones, formed in several brief incursions of the sea. On these lowland flood plains lived hordes of primitive reptiles and large amphibians, the most successful vertebrates that had yet been developed. Some of them were transitional types, intermediate between the amphibians and the reptiles. There were clumsy, short-necked reptiles (cotylosaurs) with completely roofed skulls, resembling the early amphibians. And there were more advanced, longer-necked, more slender-limbed reptiles (pelycosaurs) with an opening at the side of the skull for the jaw muscles (51). These were predatory carnivores, sharp-toothed, that ran

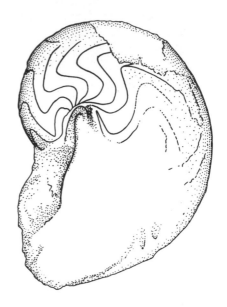

FIG. 11. Shell of nautiloid *Cosmonautilus dilleri,* from same locality as fig. 10. The simple sutures are similar to those in the shell of the living chambered nautilus (fig. 13). (From J. P. Smith, U.S. Geol. Surv. Prof. Paper 141, pl. 90.)

and jumped to seize their prey. Bizarre ones had high fringes braced by bony spines; they looked like windjammers with reefed sails (pl. 2).

Land animals of the late Permian have not been discovered in North America. Elsewhere, particularly in Russia and South Africa, mammal-like reptiles (theropsids) began to take on the characters of mammals. Several groups had mammal-like skulls, teeth, feet, and shoulder bones. These reptiles were doubtless among the most active of their kind. They were accompanied by gigantic, sluggish, herbivorous forms (late cotylosaurs) and by a few small, advanced, lizard-like types ancestral to many of the diverse dinosaurs, crocodilians, and others common in the latter days of the Age of Reptiles. The Permian in the Southwest became more arid; the reptiles are now represented only by their tracks on the wind-blown sands of the Coconino in Arizona. Finally, all trace of reptile life disappeared in North America; the climate in this latitude may have become severe.

In the Triassic, over two hundred million years ago, sea waters overran parts of northeastern and extreme southern California (fig. 12). The shores of this sea lay to the east in Utah and Arizona. Land extended thence eastward to the present coast of the Atlantic. Much of what is now Alaska, British Columbia, Oregon, Nevada, and northwestern Mexico lay under the sea.

The extent and situation of the ancient seaways are known by the present distribution of marine Triassic rocks, and the interconnections between

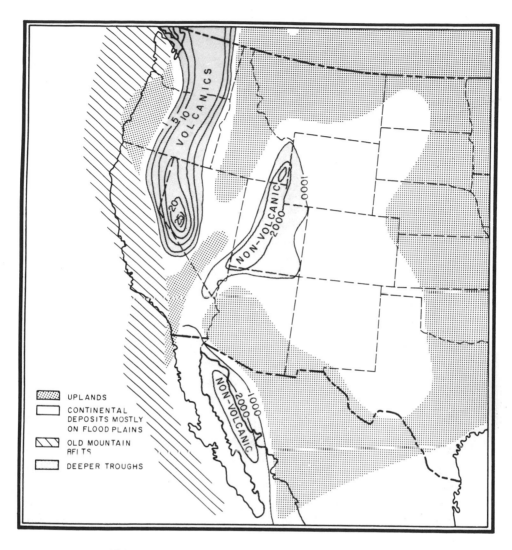

Fig. 12. Triassic seaways and basins. (After Eardley, Bull. Amer. Petrol. Geologists, 33: 674, fig. 10.)

these oceans and continental seas have been ascertained by comparison of fossil marine invertebrates, particularly mollusks. Faunas showing close relationships are considered to have lived in interconnected seas, and distant relationships imply the presence of land barriers. Ancient shore lines can be located by peculiar coastal deposits, as well as by comparison of land and marine faunas and the distribution of land faunas in beds that lie conformably under or over marine rocks—faunas that lived on flood plains near the shifting sea.

FIG. 13. Chambered or pearly nautilus, *Nautilus pompilius;* longitudinal section of shell showing large outer chamber in which animal lives, and empty chambers with internal tube (siphuncle) encasing a tenuous extension of the body. The hollow chambers serve to float the animal, which resembles the squid and octopus and like the squid propels itself by jet propulsion. (After A. K. Miller, Geol. Soc. Amer. Mem. 23, pl. 4.)

Studies of the distribution of marine invertebrates indicate that the Triassic began with a wide sea over what is now Nevada, southeastern Idaho, and northeastern California. This gulf extended northward to the Arctic. There may have been some land northwest of the California sea, for the Triassic marine faunas of Japan are not closely related to those of America. Tropical waters then extended at least as far as Oregon and supported a wealth of sun-loving animals. Over Shasta and Siskiyou counties, the Sierra Nevada, and the Inyo Mountains of today, the sea was clear and deep. Masses of clean lime were deposited on the ocean floor. Toward the south, in Arizona, great rivers flowed westward and north-westward across the flood plains.

In Triassic limestone beds, now uplifted along our mountainsides, are found the bones of marine reptiles and the coiled shells of squid-like mollusks, brilliant ammonites (fig. 10), and nautiloids with shells (fig. 11) like that of the chambered nautilus (fig. 13). And there were fringing coral reefs (fig. 14).

The most abundant of the marine reptiles were the porpoise-like ichthyosaurs. They ranged in length from two to fifty feet. Although descended from land dwellers with walking feet, they had developed the streamlined body form of fishes and porpoises, with swimming paddles, a stiff fin on top of the back, short neck, sharkline tail fins, and a long fish-catching snout armed with rows of piercing teeth. Their eyes were large, for use in deep, dark water, and the eyeball was strengthened against water pressure by a ring of overlapping bones similar to the iris diaphragm of a camera. The ichthyosaurs had no gills, and breathed through nostrils like land reptiles. They rose to the surface, as the whales do, to take deep lung-

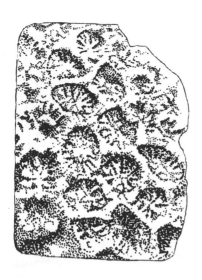

FIG. 14. Fragment of basal skeleton of reef coral, *Isastrea,* from Upper Triassic limestone, Devil Rock, Shasta Co., Calif. (From J. P. Smith, U.S. Geol. Surv. Prof. Paper 141, pl. 112, fig. 6.)

fuls of air. They were expert swimmers and did not have to crawl out on land, in the manner of sea turtles, to lay their eggs. Their young were born alive in the ocean, and possibly accompanied their parents, like young whales. The ichthyosaurs achieved complete independence of the land. They fed on fishes, principally on sharks of a peculiar type with blunt teeth similar to the teeth of the Port Jackson shark of today. And one of the California ichthyosaurs had button teeth for crushing shellfish. The ichthyosaurs were inhabitants of the world's oceans for about seventy-five million years—from the middle of the Triassic well into the Cretaceous.

The remains of another group of sea-swimming reptiles are found in the limestones of Shasta County along with the ichthyosaur skeletons and the shells of ammonites, ancient snails, and clams. These were the rare sea lizards, thalattosaurs; so far, they are known only from California. The thalattosaurs were paddle-finned, long-headed, porpoise-like reptiles, similar in appearance to the ichthyosaurs. They differed in the construction of the skull, which is more like that of the earliest lizards. The jaws swung very freely on a double hinge, for easier swallowing of food; and the snout was narrow and knifelike, armed with blunt teeth, both on the jaw and on the palate. It is possible that these reptiles fed on the smaller, thin-shelled, floating ammonites, seizing them in their hooked beaks. They were probably not as powerful swimmers and divers as the ichthyosaurs (43).

In the seas of the Triassic world there were many fishes and reptiles that have not yet been found in California but that may have lived here. The fishes and reptiles constituted the entire vertebrate population of the sea, since at this time there were no birds and no sea mammals. There were forms that would now seem outlandish.

The continental Triassic opened in the Southwest with a rejuvenation of the streams bringing gravels followed by sand and mud. This may signify an elevation of mountains in southern Arizona or northern Mexico and probably in Colorado. The shallow flood plains extend as far north as Wyoming. They were alternately flooded and exposed in sun-baked mud flats. Deposits of gypsum in the middle of the mud-flat series of rocks (Moenkopi) show that the climate was then warm and dry. Below the Painted Desert beds are the Moenkopi red sandstones and shales laid down in slow streams that flowed northwestward to the Lower Triassic embayments. Along these streams and ponds in present-day Arizona and Utah, the mud flats dried and cracked in the sun. Animals which crossed the drying mud left many tracks. These tracks were baked by the sun and then were covered and preserved by stream sands.

Such tracks are abundant at Meteor Crater and Cameron, Arizona. They represent several animals not otherwise known. One is a large reptile (*Chirotherium*) with pudgy hind feet much larger than the forefeet. The hind-foot track looks something like a hand with a "thumb" — really the little toe — turned out to the side. Along with the tracks of *Chirotherium* are the footprints of a small reptile that must have been the forefather of the dinosaur. He ran entirely on his two hind legs and used his fifth toe as a prop pointed backward to support his foot. Reptiles of this type were the first animals that learned to walk upright on the hind feet. Other tracks look like those of lizards; and there are also circular amphibian tracks which show short, blunt fingers without claws (47).

Amphibian skulls from the Lower Triassic (Moenkopi) of Arizona include forms which closely resemble the capitosaurs and cyclotosaurs of Europe, as well as the long- and slender-snouted *Aphaneramma* of Spitzbergen (67b). There is also a short, broad-headed skull, *Hadrokkosaurus*, like that of *Brachyops* from the Triassic of South Africa (67c). These amphibians indicate a land connection between the two continents, for amphibians could scarcely have traversed the open sea, since their eggs and larvae are killed by sea water.

A quiescent, unrecorded interval, Middle Triassic, was interrupted by renewed uplifts and the strewing of pebbles that now lie in thin conglomerate beds (Shinarump) over a wide area from southern Nevada, across northern Arizona and Utah, into New Mexico. Thick deposits of ash in the Painted Desert of northern Arizona were deposited upon another low flood plain (Chinle), building a sequence of fine-grained rocks of late Triassic age.

Water-loving animals — phytosaurs and giant broad-headed amphibians — inhabited the flood plains and completely died off when the flooding ceased. Dinosaurs came in the Triassic. They were small, agile, predatory, leaping and running on long hind legs, able to endure the deserts and to traverse the drifting sand.

Land life has not been discovered in the Triassic of California, because no land-laid sediments of that period have been found. On adjacent eastern shores, the low, flat, well-watered land supported tree ferns, five-fingered ferns, and other small ferns, cycads, and giant ephedras distantly related to the Mormon tea. In the early Triassic there were forests of large tree-like rushes (calamites), and these gave way to closer relatives of the modern horsetails, or scouring rushes.

The land was dotted with strands of large trees related to the modern monkey puzzle pines of Brazil and Chile. Fossil logs of these conifers are found in the "petrified forests" of Arizona, Nevada, and northern New Mexico; some of the trunks, which have turned to agate in rainbow colors, are two hundred feet long and six feet in diameter. Some of them show growth rings, indicating a seasonal rainfall, and some that had been burned on the surface show a well-preserved coating of charcoal encasing the petrified logs. Stumps and roots remain in places where they originally grew. A "stump forest" stands twenty miles northeast of the station of Adamana in Arizona. Other stumps with roots may be seen in the Blue Forest at the Petrified Forest National Monument and near Cameron, Arizona; the fossil logs there could have come from trees that grew near by. Petrified logs are now weathering from beds of sandstone, conglomerate, and shale formed in the deltas and flood plains of old rivers. In north-eastern Arizona there were great explosive volcanoes in the Upper Triassic: some of the lower shale beds consist entirely of decomposed volcanic ash deposited as mud in masses up to ninety feet in thickness.

The Painted Desert beds also contain animal bones, abundant in a few places, where fragments of carcasses were buried in the waterborne ash. There are skeletons of large, flat-headed amphibians (stegocephs), like huge salamanders, with broad bodies, short necks and legs, and short tails. They were the last of their kind (pl. 3). The stegocephs were clumsy, sluggish creatures, confined to fresh water. They moved slowly, using their weak legs to skid their awkward bodies through the reed-grown shallows. Below the chest were massive bony plates to which the coarse skin was securely attached. This equipment served as a kind of sled, particularly useful be-cause the creatures could not raise their bodies from the ground. And they needed protection from the scouring rushes, too, for those were abrasive and would wear through anything but a bone-studded hide. The mouths of the stegocephs were enormous, nearly a fourth as long and almost as wide as their bodies. The jaws and the roof of the mouth were lined with dozens of tiny, fluted teeth and two pairs of larger "tusks." The lazy creatures possibly lay quietly with their mouths open like traps, awaiting their prey —

sluggish lungfishes and other fishes. Remains of these have been found along with the stegoceph skulls.

The Painted Desert beds of Texas, Arizona, Utah, and Wyoming also yield fossil remains of land reptiles. There were light-headed dinosaurs that ran on their spindly hind legs, and stouter reptiles akin to dinosaurs, which had plates of bone over the back and spines along the sides of the body and tail. A large one, the episcoposaur, had a curved cowlike horn over each shoulder. Another, the phytosaur, had a head like that of a huge crocodile, with nostrils elevated on a crater at the summit of the head rather than at the end of the snout as in the crocodiles (12*a*).

Among the rare vegetarian reptiles was an immigrant, a big rhinoceros-like beast called *Placerias*. His ancestors were the mammal-like dicynodonts of Africa and Eurasia. *Placerias* was an aberrant monster with a pair of sharp horns pointing downward and forward from the sides of his face. He walked with his body raised on sturdy legs, and his toes were short like those of a mammal. He was one of the last of his group and became extinct toward the end of the Triassic (pl. 3).

Another vegetarian was a short-headed, rather long-necked, slender fellow *(Trilophosaurus)* with teeth set crosswise in his jaws, and with broad, high cheek bones. His remains have been found only in Texas (19). Elsewhere in the world there were many other curious reptiles, some of them ancestral to the lizards, turtles, and crocodiles; also there were rapid-running creatures, others that flew like bats, and extraordinary swimming forms with necks long and slim like fishing poles.

During the Triassic the mammals rose by slow stages from therapsid reptiles. The wonderful record of these changes is found in the rocks of South Africa. There the extensive icecaps of the early Permian gave way to a period of cool, damp climates that gradually changed into the warmer Upper Permian, when primitive reptiles lived in hordes. The African Triassic opened with a thick deposition of red beds in shallow lakes. Lime dissolved from the dolomitic highlands saturated the waters and greatly helped to preserve the bones by enclosing them in calcareous nodules. Mesozoic fossils, thus preserved in South Africa, are more abundant there than anywhere else in the world. As the African Triassic closed, aridity was widespread, and the period finally ended with the deposition of wind-blown sand in great masses. These old dunes contain the small reptiles (ictidosaurs) most closely approaching the mammals in their bony structure.

It may be that the mammals — active, warm-blooded, and covered with an insulating coat of hair — began in relatively dry regions having warm summers and cold winters, that they were arrested in their development during the more tropical periods of the late Mesozoic, and that they flourished again during the cooler, mountain-building period of the Laramide revolution at the end of the Cretaceous. It may also be suggested that the soft, sensitive skins and tactile hairs of mammals were developed in response to nocturnal habits; that the mammals worked the cool night shifts and gave way to the reptiles during the day.

The Jurassic sediments of California were laid down in seas overspreading much of the area now included in the state. The early Jurassic sea was a remnant of the great gulf of Upper Triassic time and extended over the area now occupied by the Sierra Nevada (fig. 15). Its sediments accumulated to a great depth and formed what are now the Mariposa gold-bearing slates of the Mother Lode. Collapse of this mass, and inclusion of granitic magmas within it, caused the Sierra to rise and created a new trough occupied by another seaway over what are now the Coast Ranges and parts of the Great Valley. This was the Franciscan sea of the late Jurassic (32b).

The cherts and other sediments of the Franciscan series (fig. 19) of the present Coast Ranges, far to the south and north, were then deposited to great depths until the trough sank and collapsed under the burden. This permitted subterranean magmas — serpentines and lavas — to force their way between the broken rocks and to burst out as volcanoes. Lands evidently lay fifty miles or more off the present shore, along the western margin of the Franciscan gulf, for this gulf received along its western margin coarse granitic sediments, which must have been derived from a chain of rugged mountains eroded down to its granitic core (fig. 16). The remains of these granites lie in the ocean along the continental platform today. Geologically, California extends outward for many miles beneath the sea. The offshore platform was once part of the land, and may again become a part of it (fig. 29).

Few fossils, and these poorly preserved, are known from the Franciscan rocks. Protozoans with glassy skeletons (radiolarians) have been found in the cherts, along with pieces of ichthyosaur skulls. Ammonites occur sparingly, and bits of wood (63a).

Granitic masses intruding far underground formed the present base of the Sierra; these masses began to rise, carrying on their shoulders the crumpled remains of much older rocks. Other granites were intruded beneath the Klamath and the San Bernardino mountains and mountains in the Mojave Desert. Silica, borne by hot mineralized magmatic waters, crystallized into the famous gold-bearing quartz of the Mother Lode.

The Jurassic was a period of intense development of life on the land. A new plant group, the cycads, became dominant. These palm-like plants have fronds that unroll like those of ferns and often grow from a trunk resembling a pineapple; but unlike the ferns, the cycads produce flowers. A resemblance to the pines is seen in the cone which grows at the crest of the trunk. In the cycad forests of the Jurassic were many ferns and conifers.

PLATE 3. Upper Triassic vertebrates of Arizona, in a forest of Araucarian pines partly destroyed by volcanic ash. A phytosaur *(Machaeroprosopus)*, a horned reptile *(Desmatosuchus)*, stegocephalian amphibians *(Metoposaurus)*, and dicynodont reptiles *(Placerias)*.

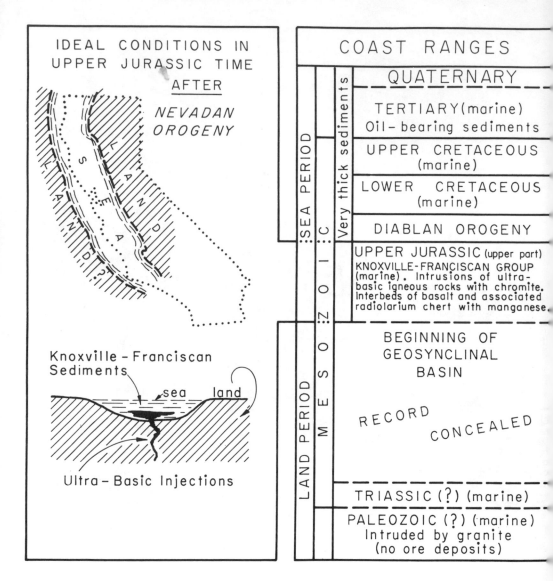

IDEAL CONDITIONS IN UPPER JURASSIC TIME AFTER NEVADAN OROGENY

LAND

SEA

Knoxville – Franciscan Sediments

sea   land

Ultra – Basic Injections

COAST RANGES

SEA PERIOD	Very thick sediments	C	QUATERNARY
			TERTIARY(marine) Oil- bearing sediments
			UPPER CRETACEOUS (marine)
			LOWER CRETACEOUS (marine)
			DIABLAN OROGENY
			UPPER JURASSIC (upper part) KNOXVILLE-FRANCISCAN GROUP (marine). Intrusions of ultra-basic igneous rocks with chromite. Interbeds of basalt and associated radiolarium chert with manganese.

M E S O Z O I C

LAND PERIOD	M	BEGINNING OF GEOSYNCLINAL BASIN
		RECORD CONCEALED
		TRIASSIC (?) (marine)
		PALEOZOIC (?) (marine) Intruded by granite (no ore deposits)

FIG. 15. Geologic events in Upper Jurassic and Cretaceous, with exte.

Cycads today, in Mexico, Cuba, and Japan, are but a remnant of the abundant forms of that period; but they are much the same, and some have trunks fifty feet or more in height (33). The late Jurassic (or possibly Lower Cretaceous) Knoxville plants of California include horsetails, ferns, cycads, and conifers—which indicates that there were damp forests here similar to those in the Lower Cretaceous of Maryland.

Fossils of ancestral crocodiles are found in rocks of the Lower Jurassic. There, too, occur the first true lizards, and a host of dinosaurs, which

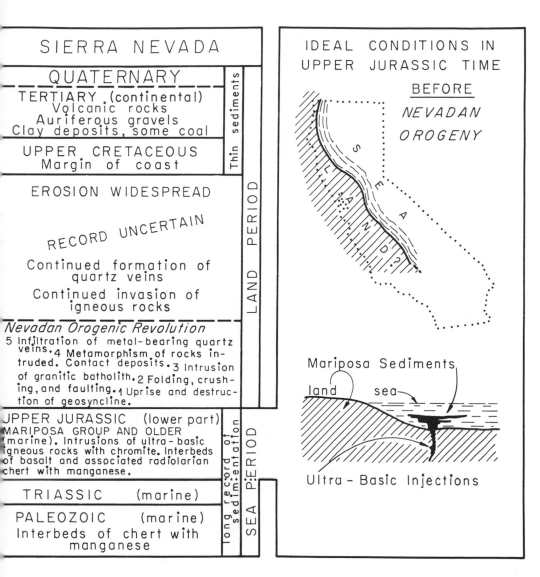

SIERRA NEVADA		IDEAL CONDITIONS IN UPPER JURASSIC TIME
QUATERNARY	Thin sediments	BEFORE *NEVADAN OROGENY*
TERTIARY (continental) Volcanic rocks Auriferous gravels Clay deposits, some coal		
UPPER CRETACEOUS Margin of coast		
EROSION WIDESPREAD	LAND PERIOD	
RECORD UNCERTAIN		
Continued formation of quartz veins		
Continued invasion of igneous rocks		
Nevadan Orogenic Revolution 5 Infiltration of metal-bearing quartz veins. 4 Metamorphism of rocks intruded. Contact deposits. 3 Intrusion of granitic batholith. 2 Folding, crushing, and faulting. 1 Uprise and destruction of geosyncline.		Mariposa Sediments land sea
UPPER JURASSIC (lower part) MARIPOSA GROUP AND OLDER (marine). Intrusions of ultra-basic igneous rocks with chromite. Interbeds of basalt and associated radiolarian chert with manganese.	long record of sedimentation	SEA PERIOD
TRIASSIC (marine)		
PALEOZOIC (marine) Interbeds of chert with manganese		Ultra-Basic Injections

...efore and after uplift of Sierra Nevada. (From Jenkins *et al.*, 32*b*, map.)

toward the end of the period developed into the largest land animals that ever lived. These were the huge sauropods, *Brontosaurus* and *Diplodocus,* swamp dwellers; the brachiosaurs of East Africa, largest of all, reaching a height of forty feet; and the plated stegosaurs, common at Dinosaur Monument in northeastern Utah (pl. 4). Bipedal dinosaurs of the more agile, sand-dwelling types inhabited the deserts of the Southwest. Their bones and tracks lie in the desert sandstones of the Wingate and Navajo formations of Arizona and Utah. Some of the desert dinosaurs laid eggs; their fossil "nests"—clutches of petrified eggs—have been found lying in the Jurassic sands of Mongolia. A number of small, batlike, flying reptiles (pterodactyls) have left impressions of their flight membranes, and feathered birds *(Archaeopteryx)* have made famous the Solenhofen limestone quarries of Bavaria. These were long-tailed, toothed, reptilelike birds with claws at the front margin of the wings. They undoubtedly could fly short distances, and had perching feet. The rare mammals, barely started on their career, were small, weak-brained forms (51).

Seagoing plesiosaurs and ichthyosaurs, more advanced than those of the Triassic, were widespread. Archaic fishes (crossopts) approached the end of their span. They did not appear later, except for a remarkable living individual recently fished up in the Indian Ocean off the coast of South Africa. Sharks were numerous. A host of porcelain-scaled fishes (ganoids) inhabited the fresh waters. And the modern bony fishes allied to the carp and goldfish made their first appearance in the late Jurassic. These are the ancestors of most of the fishes of today.

Dominant in the seas were the floating coiled ammonites—creatures which have left varied shells in such abundance that the age of marine rocks may be accurately determined by studying them. Their distant relative, the chambered nautilus, rare today, looks like an octopus with a shell on its back. Like the nautilus, the ammonite lived only in the terminal, outer chamber of its spiral shell, adding chambers as it grew, and moving out of the old chamber into the new one. The chambers are locked together by sutures in complex patterns that are used by stratigraphers to correlate and date the rocks.

Hosts of floating and fixed stone lilies (crinoids) with their feathery arms encircling the mouth—bulbs on stalks and flexible stalks formed of perforated discs—swarmed in the Jurassic seas. They are wonderfully well preserved in the slate quarries at Holzmaden, Germany. Brachiopods, plentiful in the Paleozoic, were much fewer in the Jurassic. The modern lobsters, crabs, and crayfish appeared. Modern types of insects occurred for the first time. Reef corals and sponges were numerous and varied.

The Cretaceous witnessed a long, complex series of events; its close, some seventy million years ago, brought the Age of Reptiles to a tragic end. The Cretaceous opened with swamps and coal deposits over much of North America. The old Permian mountains had been continuously eroded, and but few highlands remained, except in the West where the Sierra had begun to wear away. The sea encroached on the land in many places. Coming in at first from Mexico, it formed a gulf that extended through Texas and Louisiana. This influx of the sea was interrupted three times by uplifts. A long seaway, the Tethys, crossed southern Europe and Asia, where the Alps, the Caucasus, and the Himalayas now stand. The low-lying

Atlantic seaboard and the great valley of California were constantly under water and received heavy loads of sediment. Toward the end of the Cretaceous a great flooding occurred, particularly along the trough that extended from the Gulf of Mexico to the Arctic by way of Kansas and Nebraska.

This inland sea nourished swarms of microscopic foraminifers. Shells of the forams produced the great chalk beds. Swimming in the warm, shallow waters were giant predatory fishes, ichthyosaurs, plesiosaurs, and large porpoise-like lizards, the mosasaurs. Flying and diving over this sea were huge soaring reptiles *(Pteranodon)* with wings that extended more than twenty feet, and with long, toothless, fish-catching beaks compensated on the back of the head by a broad, balancing blade like a weather vane — doubtless to keep the head pointed in the direction of flight (pl. 5). Waterfowl, equipped with teeth, and basking turtles swam in this great sea.

Some of the coiled ammonites grew to mammoth size, reached their climax, and at the end of the Cretaceous became extinct. The plesiosaurs and mosasaurs also died out. Ichthyosaurs dwindled away. The dinosaurs, extraordinarily profuse and varied toward the close of the period, became extinct at its very end.

The ancient sauropod dinosaurs had lingered into the Lower Cretaceous of North America and survived in abundance south of the Equator. Their place in the North was taken by the bizarre duckbills, the horned and armored forms, and the semiaquatic swamp swimmers. The harmless herbivores were preyed upon by carnivores, including the awesome *Tyrannosaurus rex,* whose huge skull was armed with a whole battery of teeth like bayonets.

The incredible world of reptiles came to a climax and mysteriously ebbed during the Cretaceous. A few forms persisted into the Tertiary: the crocodiles, turtles, lizards, and early serpents, which had barely started. Archaic mammals, marsupials of the opossum type, and primitive insectivores went on through into the Eocene. And there were many invertebrates that continued on land and sea.

The extinction of the large reptiles at the close of the Cretaceous is one of the startling events of paleontology. Geologists point to the Laramide uplift, a revolutionary earth movement which occurred at the end of the Cretaceous, when the great inland seas retreated and the swamps were for the most part drained. But this revolution was not world-wide; there is no evidence of it in California, where the marine sequence of rocks is continuous into the Paleocene.

There was no great obliteration of plant life. Flowering types such as magnolias, palms, figs, alligator pears, willows, camphor trees, and oaks first appeared in the Cretaceous and still exist today. This makes the extinction of animal life even more puzzling. Certainly there was no period of

PLATE 4. Upper Jurassic life in northern Utah (Dinosaur Quarry National Monument). A long-necked dinosaur *(Diplodocus)* wades near a swampy shore, where a carnivorous dinosaur *(Ceratosaurus)* feeds on one of the vegetarian plated dinosaurs *(Stegosaurus)*. Cycads, ferns, horsetails, and in the distance, tree-ferns.

PLATE 5. Upper Cretaceous reptiles of California, from the Moreno formation, west side of San Joaquin Valley. A long-necked plesiosaur *(Elasmosaurus)* and a mosasaur *(Plotosaurus)* swim near cliffs where flying dragons *(Pteranodon)* swoop and glide. A duck-billed dinosaur *(Prosaurolophus)* forages along the shore.

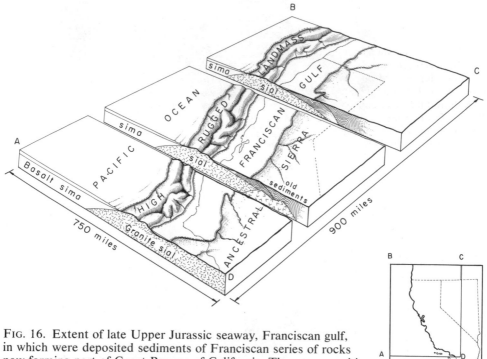

FIG. 16. Extent of late Upper Jurassic seaway, Franciscan gulf, in which were deposited sediments of Franciscan series of rocks now forming part of Coast Ranges of California. The worn granitic ridges of the offshore lands contributed recognizable sediments to the Franciscan gulf. This is the main evidence of their presence. (After Taliaferro, 63b, fig. 3.)

extreme cold, or tender Cretaceous plants would have perished. According to H. C. Urey's data, the temperature of the North American inland sea increased to about 26°C. during the mid-Cretaceous and cooled toward the end; therefore the extermination of the sea reptiles could scarcely have been caused by excessive heat. The rise of land mammals may have crowded out the less adaptable reptiles but would hardly have affected the sea reptiles and the ammonites. Nevertheless, the world of animal life, slowly developed during the Age of Reptiles, was decimated, and the modern world began to take form.

California seas lay farther east during the Cretaceous than they did immediately before it (fig. 20). The Sierra, the Klamath region, and nearly all of southern California were above sea level. A peninsula may have extended out to sea northwest of San Francisco, and the present Channel Islands formed the center of a land mass southwest of Los Angeles. The eastern margin of an inland sea lay west of Fresno, and along this shore roamed duck-billed dinosaurs (*Prosaurolophus*) bones of which have only recently been discovered. Paddling in this sea were porpoise-like sea

Fig. 17. Glassy "shell" (cell capsule) of diatom, *Arachnoidiscus*, from diatomite deposit (Monterey Miocene) at Lompoc, Calif. The thick diatomite beds consist principally of broken fragments of many kinds of diatoms and radiolarians. (After Henry Mulryan, Calif. Jour. Mines and Geol., Report 32 of State Mineralogist, photo no. 5.)

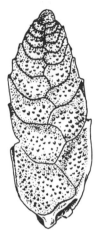

Fig. 18. Shell of foraminifer *Bolivina bramlettei,* from Upper Miocene diatomite, Palos Verdes Hills, Calif. The shell is perforated with small pores (foramina) through which protruded streaming threads of protoplasm from the living cell body within the shell. The substance of the shell is chalky lime. (After R. M. Kleinpell, *Miocene Stratigraphy of California*, Tulsa, 1938, pl. 21, fig. 10.)

lizards (mosasaurs) (12*b*), and plesiosaurs (67*a*) with diminutive heads on the end of snaky necks fifteen feet long (pl. 5). Small saw-toothed sharks and other fishes were numerous. Bushy hydrozoans grew in the shallow waters. Microscopic plants (diatoms) with delicate glassy shells like pillboxes appeared for the first time (fig. 17). Foraminifera, the chalk-forming protozoans, became abundant (fig. 18). Cretaceous rocks, deposited in the California seaways, appear now as upraised ridges along the west side of the San Joaquin Valley, as well as throughout northern California and near Los Angeles.

Climates during the Cretaceous were remarkably uniform, warm, and mild over North America. Remains of tender plants have been found even in northern Canada. The period saw the climax and decline of the ammonites and reptiles, the beginnings of the active placental mammals, toothed sea fowl, flowering plants, and social insects. Perhaps the first music in this ancient world was the stridulation of the insects; the first dance, that of the two-legged dinosaurs; the first architecture, the nests of the termites and the birds.

Transformation

Two million centuries ago the reptile hordes found tempting foods among the new-formed plants: juicy stems and roots, leaves, buds, and pods. Thus, reptile herbivores gained special grinding teeth and horny, toothless beaks, like those of tortoises. Plant food was plentiful: plant eaters grew in numbers and diversity, beyond all that had gone before.

Meanwhile, predacious reptiles stalked their prey among the herbivores, and grew more adept, quicker limbed and lighter footed, with firmly rooted teeth—teeth to seize and kill, and teeth to cut and rend and crush the flesh and tasty fat. They chewed with shortened jaws; and bony-plated palates grew between the chambers of the mouth and nose, preventing strangulation by the food. Along with shortening of the jaws came alterations of the bones around the ear: transfer of disused jawbones to the middle ear, and shifting of the drum beyond the swelling muscles of the jaws.

In the Triassic, carnivores and herbivores, the preyers and the prey, gained speed and cunning, brought the limbs beneath the body by turning in the elbows and the knees, rotated the feet to best direct the shortened toes, and raised the body from the ground. Such a stance required endurance and agility. The brain kept step, swelled fore and aft, giving new circuits for control of balance and recording of experience. The blood machinery kept step, with separate chambers in the pumping heart and re-division of the arteries. The body chemistry kept step. with regulation of the heat supply. Warm-bloodedness required an insulated skin; hence, tough reptilian skins grew soft and furry. Fur turns brittle without oil, and so the skin was pocketed with lubricating glands of oil.

With all this new equipment, habits changed. The nest-born young snuggled for warmth against the body of the mother, who licked and tended them. They too licked the skin and hair of the mother's breast. Secretions of the skin at last poured out as milk, first to be lapped from folds and pouches, then sucked when teated breasts were formed.

Thus came the transformation, during the long years of the Jurassic, of mammal-reptiles into real mammals, and the rise of mother love and family care.

CHAPTER 7

Age of Mammals

THE MAMMALS PLAYED MINOR ROLES in the drama of the Age of Reptiles. After the curtain fell at the end of the reptile act, the mammals burst upon the stage in the Paleocene and Eocene. Mammal monsters took over in the sea where reptile monsters swam: whales replaced the sea reptiles. Bats, instead of pterodactyls, flitted through the air. Birds swam, dived, soared and fluttered, and began to sing. The land teemed with a host of ambling, running, galloping, leaping, burrowing, climbing, and gliding mammals that fed on seeds, fruits, stems, leaves, roots, insects, worms, crabs, fish, frogs, and the flesh of their own kind. The warm-blooded mammals maintained themselves in cool climates, where reptiles became torpid. Mammals inhabited the North, where reptiles never penetrated. Some of the mammals increased in size and became cumbrous in body like the elephant, rhinoceros, and hippopotamus of today. Others became light-limbed and marvelously swift of foot.

The early mammals were small-brained in comparison with their descendants. Only a few of the primitive ones still survive — relatives of the opossums, hedgehogs, and moles — relics on isolated islands and in tropical forests. Among the early ones were the ancestors of modern horses, tapirs, and rhinos (perissodactyls); of the cattle, deer, giraffes, camels, pigs, and hippos (artiodactyls); of the squirrels, beavers, porcupines, guinea pigs, chinchillas, mice, rats, and jerboas (rodents); of the elephants (proboscideans); of the hyraces and sea cows and whales; of the pangolins, aardvarks, sloths, and giant anteaters (edentates); of the seals and walruses, wolves, bears, raccoons, pandas, wolverines and weasels, civets, hyenas, and cats (carnivores); of the tree shrews, lemurs, monkeys, apes, and man (primates). Our modern circus parade trails back into the Paleocene. The evolution of such forms as elephants, camels, and horses is well recorded by the fossils and indicates a slow and not too steady progression of changing types through sixty million years or more (20).

Mammal evolution is a well-documented fact of history. Its relations to early man are significant, but that story is not yet fully told. There is much to be learned from the fossils; but no early human fossils have been found in the Western Hemisphere, for man came here late in his history at a time to be measured in thousands rather than in hundreds of thousands of years.

Mammalian evolution is the best known of all the great fossil stories. The fossils indicate a series of supreme attempts delicately to adjust habits, physiology, and bodily structure to the varied surroundings and conditions under which life may exist. It provides examples of the growth and flowering, as well as the decay and death, of many honored family trees.

The mammals struck out in all directions to find new ways of living. Some went into the ground to become burrowers like the ground squirrels, or earth dwellers like the gophers, in which the ears, eyes, and tail have become reduced. The permanently subterranean moles have lost the external ears and eyes. Moles swim through the earth, using their enormously strong arms and broad forepaws as paddles, turning first on one side and then on the other to raise the roofs of their tunnels. The burrowing marsupials of Australia, though not related to the moles, include subterranean forms that resemble the golden moles in many curious details. Mammals have developed hopping types, such as the kangaroo rat and the jerboa; the independently developed wallabies of isolated Australia parallel these. This again shows how closely some anatomical mechanisms are tied to a special way of life: how closely the course of evolution is channeled into the environment.

The body forms of porpoises, ichthyosaurs, and the marlins are remarkably alike; yet the internal parts show that the porpoise is a mammal, the ichthyosaur was a reptile, and the marlin is a fish. All three were developed independently. The porpoises and ichthyosaurs had land-living, four-footed ancestors. The flying phalangers of Australia are astonishingly like the flying squirrels of other parts of the world; yet no close relationship exists, for one is a marsupial, the other a placental. The litopterns of the ancient fauna of South America resembled the horses of the northern world, but were not related to them. The saber-toothed marsupials of the South American early Tertiary were duplicated by the saber-toothed cats that replaced them when that continent gained a connection with the North (51).

When this connection was established, a horde of strange beasts crossed it: the small armadillos and the giant glyptodonts, the large and small ground sloths, the Canada porcupine. In exchange, a host of North American animals drifted south: the bears, weasels, wolves and cats, the deer, horses, tapirs, llamas, elephants, and mastodonts.

California received a share of the emigrants from South America. Some, like the ponderous ground sloths, dominated her landscapes until a few thousand years ago. Clumsy, turtle-backed glyptodonts came for a short visit before they were extirpated. The porcupines still nibble bark from the trees of our forests.

When the Age of Mammals opened in the early Tertiary, western California lay beneath the sea. The southern and northern areas, the Sierra, and the deserts were high ground (fig. 21). In Oligocene time, the sea had retreated, leaving only one large embayment—in the San Joaquin Valley (fig. 22). Twenty million years ago, in the middle Tertiary (Miocene), seas extended over the San Francisco Bay region, the lower San Joaquin Valley, and coastal southern California (figs. 23–25). The Pliocene, about five million years ago, was a period of recession, which ended with fewer embayments and more high ground than there had been previously. The San Joaquin seaway was restricted to the Bakersfield-Coalinga area. Paso Robles, Santa Maria, Ventura, and Los Angeles were still covered. The Colorado Desert was inundated. Lands extended into the ocean off what is now the central coast (fig. 26). In the Quaternary, approximately the last million years, the land rose to its present position except for fluctuations of sea level due to the accumulation and melting of ice in the North (48).

Life in these seas, with minor differences due primarily to differences in temperature, was much like that of today. Large orbitoid foraminifers swarmed in the warm inland waters (fig. 27). Ornamented shells, now found in the tropics, abounded in the early and middle Tertiary seas of California. Some of these are related to those found in Japan, some to those found in the Caribbean today. Cooler seas prevailed in the later Tertiary, and then in the late Pliocene came cold-water faunas which have continued to the present. With some fluctuation, there was a moderate cooling of California coastal waters from the Tertiary to the Ice Ages of the Pleistocene.

The shellfish reached a peak of development in California in the Upper Miocene, when oysters were giants, up to eighteen inches in length, with shells more than an inch thick. Huge barnacles *(Tamiosoma)* clustered on the sea floor in crowded colonies, formed tubular shells like great honeycombs, and lived in the top story of each cell.

Microscopic diatoms with glassy shells (fig. 17) swarmed in clear, quiet bays in such profusion as to form thick white deposits, like those at Lompoc where millions of cubic feet of pure diatomite are now being quarried for industrial use. Tiny plants and other creatures lived by the quadrillions in the coastal waters, and their skeletons contributed to the formation of vast layers of rock in the Coast Ranges. Although the pearl oyster, *Margaritana,* does not now live as far north as California, its remains are found here, and fossil pearls have been discovered within the shells. Scallops *(Pecten)*, clams, mussels, cockles, whelks, nautiloids (fig. 28), abalones, sand dollars, sea urchins, crabs, and all the varied life of the modern sea prevailed in even greater variety in the waters of California during the Age of Mammals.

California, in the early Tertiary, was a subtropical region, warm and rainy, like Florida. Plants now extinct grew in the lowlands, along with more familiar recent types: the walnut, the magnolia, the fig, and the palm. Coal beds were laid down where Mount Diablo now stands.

A long procession of strange beasts lived and died and left their bones in a few places in the rocks. Knowledge of the earliest ones comes from the Sespe beds lying chiefly in Ventura County, and from the Poway conglomerates near San Diego, The Sespe fossils show that the mammals resembled in part those from the Upper Eocene of northern Utah (Uinta) and those of the Upper Oligocene of Oregon (pl. 6). There were the great-grandfathers of the dogs and cats (creodonts), light-limbed running rhinos *(Triplopus),* and ponderous relatives of the rhinos (titanotheres) remains of which have also been found at Titus Canyon in Death Valley (pl. 7). Squirrel-like ancestors of the modern tarsiers climbed the trees of the rain forests. *Chumashius* (named for the Chumash Indians) was one of these. He was close to being an ancestor of the lemurs, monkeys, and man (pl. 6). Surely there were hosts of others, running, digging, climbing, and swimming in the streams and seas; but not many have been found among the scanty fossils from the California Eocene (62c).

The Rockies and high plains from New Mexico to Montana contain fossils which show how the horse began as a small creature *(Eohippus)* with four hoofs on each front foot and three on each hind foot, blunt squirrel-like teeth, and a short muzzle. The largest species of *Eohippus* was about the size of a small sheep. The monstrous walking bird, *Diatryma,* could have killed one of them with a single blow of its great beak. During the Tertiary, the horses gradually grew larger; the feet became longer and lost the side toes; the teeth increased in height and in number of enamel folds, for heavy duty; the muzzle became longer and the brain larger. And in the early Pleistocene the horses were already like those of today. Remains of the Oligocene *Miohippus* have been found in the upper Sespe beds of California along with gracefully built, slender-limbed, small rhinos, piglike deer (oreodonts), an early type of camel, primitive carnivores, and a rodent or two.

The closest living relative of those early rodents is the burrowing

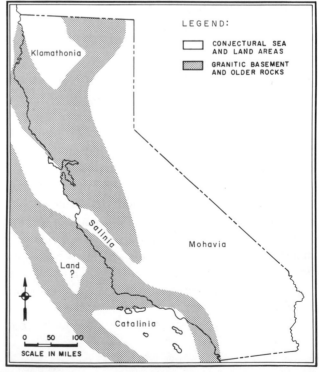

FIG. 19. Present extent of Franciscan series; thicknesses of sediments indicated by contours. (Principally after Reed, 48, fig. 55.)

LEGEND:

post-JURASSIC SEDIMENTS

PRESENT EXTENT OF FRANCISCAN SERIES

LAVA BEDS

GRANITIC BASEMENT AND OLDER ROCKS

SAN ANDREAS RIFT

SCALE IN MILES
0   50   100

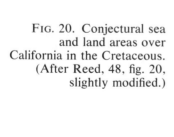

FIG. 20. Conjectural sea and land areas over California in the Cretaceous. (After Reed, 48, fig. 20, slightly modified.)

LEGEND:

CONJECTURAL SEA AND LAND AREAS

GRANITIC BASEMENT AND OLDER ROCKS

Klamathonia

Salinia

Land ?

Mohavia

Catalinia

SCALE IN MILES
0   50   100

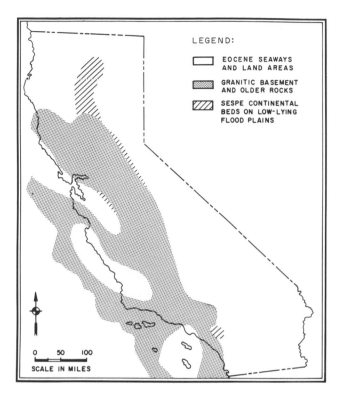

LEGEND:

☐ EOCENE SEAWAYS AND LAND AREAS

▨ GRANITIC BASEMENT AND OLDER ROCKS

▧ SESPE CONTINENTAL BEDS ON LOW-LYING FLOOD PLAINS

0   50   100
SCALE IN MILES

FIG. 21. Eocene seaways and land areas. (After Reed, 48, fig. 26.)

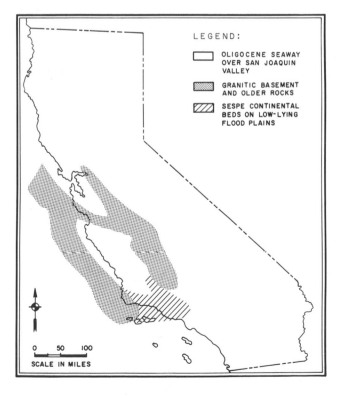

LEGEND:

☐ OLIGOCENE SEAWAY OVER SAN JOAQUIN VALLEY

▨ GRANITIC BASEMENT AND OLDER ROCKS

▧ SESPE CONTINENTAL BEDS ON LOW-LYING FLOOD PLAINS

0   50   100
SCALE IN MILES

FIG. 22. Oligocene seaway over San Joaquin Valley region. (After Reed, 48, fig. 27.)

89

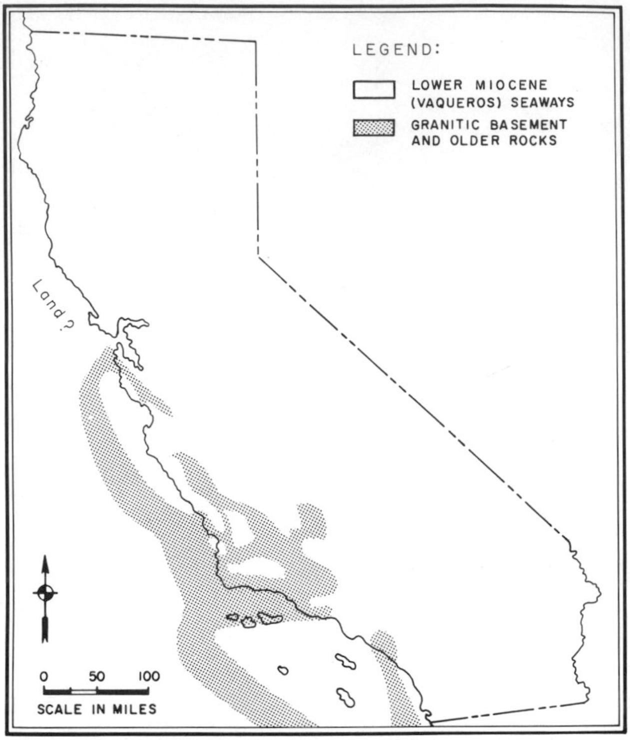

FIG. 23. Lower Miocene seaways. (From Reed, 48, fig. 31.)

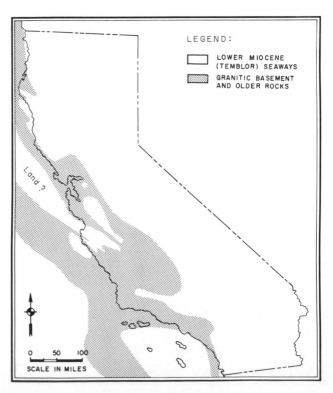

FIG. 24. Lower Miocene seaways. (From Reed, 48, fig. 28.)

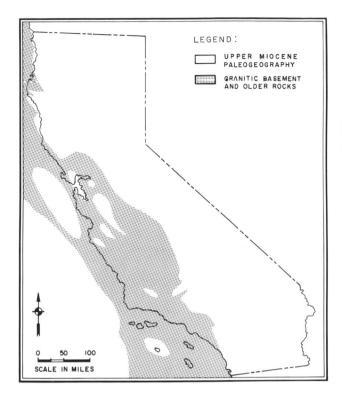

FIG. 25. Upper Miocene paleogeography. Some of the "Upper Miocene" deposits are placed in the Lower Pliocene by some authors. (From Reed, 48, fig. 42.)

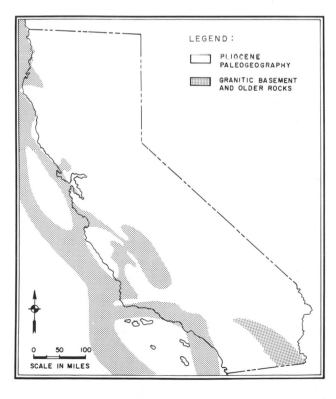

FIG. 26. Pliocene paleo-geography. In the Lower Pliocene (Orindan) there was evidently a granitic land mass connected to the mainland near San Francisco, the source of eastward-flowing streams. (Adapted from Reed, 48, fig. 51, slightly modified.)

91

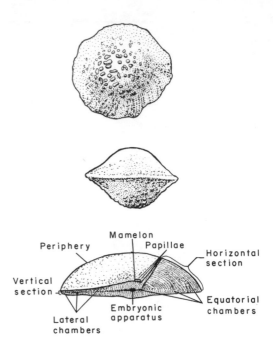

FIG. 27. *Discocyclina californica,* large orbitoid foraminifers, restricted to the Middle and Upper Eocene of California. (After H. G. Schenck, Trans. San Diego Soc. Nat. Hist., 5, fig. 3, and pl. 27, figs. 4-6.)

Aplodontia, or sewellel, found in the wet ground around springs and streams at Point Reyes, Point Arena, in the Sierran meadows above Yosemite Valley, and in the damp forests of Humboldt and Siskiyou counties. *Aplodontia* is a relic of the past, barely able to survive now in isolated colonies where soft-barked shrubs, ferns, and succulent herbage are available for its poorly developed teeth. Its fossil remains in the desert Miocene rocks of northern Nevada and eastern Oregon show that this now-dry region once supported lush vegetation. The Cascades arose and cut off the moisture-laden winds that once brought abundant rains to those eastern regions.

The Lower Miocene witnessed a significant event: the rise of the cereals, harsh grasses of the plains. Their tough seed husks have been found, silicified and well preserved, in the sandy rocks. The grasses provided a new source of abundant, concentrated food. But to eat this required a battery of good, strong, long-lasting teeth. The ancestral horses, cattle, antelopes, and camels that came out of the swamps and forests to live on the plains developed into longer-footed, swifter-running forms, with teeth which grew almost as fast as they were worn down. Lingering in the forests were the beavers, tapirs, rhinos, titanotheres, pig deer, and mastodonts; these became scarcer, moved away, or changed their habits as the forest dwindled and retreated. To the plains also came the newly developed wolves and

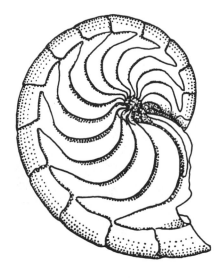

Fig. 28. Internal mold of shell of Tertiary nautiloid from the Temblor Miocene near Bakersfield, Calif. The sutures are similar to those of the living chambered nautilus (fig. 13). (After Miller, Geol. Soc. Amer. Mem. 23, pl. 91, fig. 2.)

swift-running cats with feet strengthened to catch the galloping herbivores.

These world-wide changes were shared by the land life of California. The youngest beds of the Sespe and the deposits in the hills at the south end of the San Joaquin Valley yield evidence that the older forest faunas still lingered; and the Barstow Miocene on the Mojave, laid down later, contains the remains of the camels and antelope horses of the early plains. Along with them were rabbits, large land tortoises, aberrant antelopes with horns on their noses *(Syndyoceras),* and the New World pigs called peccaries. Seals, sea lions, a varied lot of porpoises and whales, and sea cows became abundant. Along the coasts of California and Japan lived some peculiar beach-dwelling monsters (desmostylids): some looked like elephants; others, more like sea cows.

In the Pliocene, plains animals became dominant in California, and more diverse. Extraordinary, long-necked giraffe camels, large, single-hoofed horses *(Pliohippus)* hyena-like dogs, as well as huge bear dogs, saber-toothed cats, ground sloths upon which the saber cats feasted, huge lion-like cats, deer and antelopes with marvelously developed horns and antlers, big land tortoises, and small beavers comprised the life of the Pliocene. Elephantine mastodonts, with long tusks in their lower jaws, dredged the more succulent vegetation along the streams and in the ponds. Their heads were long; their legs, and probably also their trunks, were

PLATE 6. Early Tertiary mammals of California, from the Sespe beds at Pearson Ranch, Ventura County, Upper Eocene (see Stock, 62c). Early tarsioid lemur *(Chumashius)*, ancestral creodont carnivore *(Hyaenodon)*, titanotheres *(Teleodus)*, primitive rhinoceros *(Amynodontopsis)*, and running rhinoceros *(Triplopus)*.

Plate 7. Landscape in early Oligocene at Titus Canyon, near Death Valley (see Stock, 62c). Large titanotheres *(Protitanops)*, and early three-toed horses *(Mesohippus)*.

short. They must have moved in herds of hundreds, for in the Black Hawk quarry their bones have been found in masses.

The Black Hawk quarry is one of the rare sites containing mammals of the California Pliocene all buried together in the ash, gravel, and clay of an old stream channel. It lies on one of the southernmost foothills of Mount Diablo, where it was discovered by the paleontologist Bruce L. Clark, who noticed fragments of bones and teeth lying in the grass. A little digging disclosed hundreds of fossils in the soil; the best-preserved of these were the heavy teeth of mastodonts, *Gomphotherium* (see jacket and end papers). Because the fossiliferous layers, originally nearly horizontal, had been tilted back at a steep angle into the hill, it was necessary, after the fossils were gathered at the surface, to cut a deep trench to expose those below. Fossil leaves show that in the Pliocene the Black Hawk landscape was dotted with groves of manzanitas, oaks, and sycamores. The land was flat, except for the distant Sierra with its volcanoes and lava flows. The San Joaquin Valley was an embayment.

The Black Hawk animals were those of the plains and the river courses (38). There were ancient rabbits, ground squirrels, extinct beavers *(Eucastor),* a short-headed, bone-chewing dog *(Osteoborus),* a big bearlike dog *(Aelurodon),* a fox, weasels, raccoons, ring-tailed cats *(Bassariscus),* a large lionlike cat *(Pseudaelurus),* a small saber cat, herds of mastodonts, horses of two kinds, a now-extinct peccary, two kinds of long-necked camels, a pig deer (oreodont), and a deerlike, hornless antelope *(Merycodus).*

Volcanoes were widespread in northern and central California, from the Pliocene almost to the present time. Volcanic ash near Calistoga at the head of Napa Valley has buried and preserved the great trunks of redwoods and other trees. In northern Nevada, ash-buried tree trunks have been turned into precious opal, and fossil bones of rhinos, horses, and camels are found there too. At Rodeo, on the south shore of San Pablo Bay, the ash has covered bones of rhinos, the first single-toed horse, the last of the slender-limbed three-toed horses, wolves, cats, antelopes, and frogs.

The Berkeley Hills are crowned with remnants of ash beds and lava flows which lie above a pebbly conglomerate laid down by streams that swept across this region from granitic mountains beyond the Golden Gate, before the Berkeley Hills began to rise. Fossils from this conglomerate and from thin, fresh-water limestone beds above it, show that ancient beavers, geese, camels, old-time peccaries, ancestors of the pronghorn antelope, three-toed horses, and bear dogs occupied this region during the Pliocene (61). Death Valley's Furnace Creek formation contains the tracks of Pliocene horses, dogs, and small mastodonts, as well as fossil leaves (46).

Late Pliocene animals such as coyotes, bears, cats, raccoons, horses, and camels looked much like those of today. The rhinos became extinct over the whole Western Hemisphere. Ground sloths and giant armadillos (glyptodonts) entered from South America. The pig deer disappeared, also the three-toed horses (*Hipparion*).

The Pliocene ended with the advance of cold climates and the building up of masses of perpetual snow destined to form icecaps over Canada, the Great Lakes, and the Sierra Nevada.

Deep Freeze

The Pleistocene snow falls softly, covering the grass and bushes on the hillsides, enveloping the trees, and drifting across the valleys. In the mountains, the firs, limber pines, and aspens bend beneath the load and bury their heads and branches under the snowbanks. Deer and mountain sheep, followed by cougars, coyotes, and wolves, straggle toward the lowlands, driven by the storms. Bison and antelope on the plains turn southward, bunching up at night in draws and hollows with their tails to the icy wind. Elephants, horses, and camels follow dwindling pastures to warmer climes. Bears, martens, fishers, wolverines, pikas, lemmings, chipmunks, and squirrels remain in the mountains — holed up in their burrows, caves, and hollow logs.

A cool sun melts the surface snow. Cold nights freeze the melted film, and the snow pack settles firm. Storms come again, and the drifts pile higher. Even on level ground the frozen snow reaches far above the lower branches of the pines — a white blanket covering the land.

Spring comes, and the snow lies deep. Summer, and the melt-water rushes off to flood the rivers, but some snow stays. Autumn, and cool nights freeze the drifts still bedded in the deep ravines.

So it goes, year after year. For ten centuries or more the climate becomes cooler: cold, wet winters, cool summers, then a few warmer years when the bent trees rise like Old Rip from their long sleep beneath the snow. Finally comes a time when the meager summer sun, shut off by banks of alpine clouds, no longer penetrates to melt the snow. Life has left the highlands or is buried forever beneath the drifts. Only a few hardy birds hop about the glaciers, seeking insects blown up from the lowlands and frozen stiff.

The highlands refrigerate and condense the moisture in the air and blanket themselves with cool summer fog. The meager summer melt permits each winter's snow to pile above what came the year before. The great mass turns to ice and starts to grind, slowly crunching its way through the gorges, plucking masses of rock from the mountainsides and gouging deep hollows in the canyons. Across gentle slopes and over ridges the ice heaves and jerks. Over the rock pavements, hummocks, and domes, the granite is polished smooth, pressed hard, grooved and scratched and chatter-marked by boulders embedded in the creeping ice.

CHAPTER 8

Coming of the Ice

I MAGINE YOSEMITE in the last ice period of the Pleistocene. The valley is filled with a river of ice, level with the top of Lower Yosemite Fall. The glacier lies silent and white as far down as El Capitan. Along its sides are piled long ridges of broken rock from the cliffs above. The moving ice carries this debris as on a slow conveyer belt, and dumps it where the glacier melts at its moraine; thus millions of tons of good Sierran granite are transported from the heights to the foothills. In the center of this slowly flowing ice is another long ridge, or median moraine. Below Half Dome is the junction of two glaciers—one from the east and the other from the north. They join to form the main Yosemite Glacier. One comes through Tenaya Gorge and the other from Little Yosemite over thundering ice falls at Nevada Fall and Vernal Fall. Each of these tributary glaciers bears a lateral moraine, or ridge of debris, on each flank. Where the two join, the adjacent moraines combine to form the central ridge of the big glacier. The High Sierra is heavily clothed in ice. Tuolumne Meadows, Lake Tenaya, all the region around Mount Lyell, and Little Yosemite are deep under the glaciers. Only the highest summits show as dark spots above the expanse of white.

During one early glacial stage of the Pleistocene, the ice lay two thousand feet deep in Yosemite Valley and crept down as far as El Portal before it melted. The domes were then covered, and only the highest summits of Lyell and Dana were ice free. The floor of Yosemite was gouged hundreds of feet below its present level and, when the ice melted, held a picturesque lake 2,000 feet deep.[1] The river brought silt and sand, which finally filled this lake and produced the flat meadows of the valley floor as they exist today (41).

Four times at least the highlands lay beneath the ice. Each time, the canyons and lake beds were deepened, the canyon walls were polished, drifted boulders and sand were piled in rough moraines, and the rivers were clouded with rock flour powdered by the grinding ice. After each advance of ice came climates mild and rainy, hot and dry, to last an interval before

---

1. Yosemite Nature Notes, vol. 35, no. 1, p. 5.

the next onslaught of cold (52b). The Sierran ice stages have been named the McGee, the Sherwin, the Tahoe, and the Tioga—in that order from older to younger. Remnants of the two earlier moraines lie high above the present canyons on the Sierran east slope; the boulders they contain are decayed. But the smooth pavements and boulders left by the Tioga ice are nearly as fresh today as when they were formed, a mere ten thousand years ago. The California mountain glacial advances may be equivalent to the stages of the great ice sheets of the East—the Nebraskan, Kansan, Illinoian, and Wisconsin—but the correlations are still in doubt (7).

Glaciers and ice fields in the Pleistocene lay deep in the High Sierra, and a small mass of ice accumulated on the north side of San Gorgonio Peak in the San Bernardino Mountains, far to the south. Meltwater flowed across the deserts and dotted them with lakes, the shimmering salt flats and playas of today. The chain of desert basins from Owens Lake through Searles, Panamint Valley, and on into Death Valley—a distance of two hundred miles—formed a long, connected waterway. The outlet from Owens Lake ran southward through Haiwee. Here the river had rapids and falls, and its channel may still be clearly seen. The old river emptied into Searles Lake, now a salt and borax pan on the desert. From there it overflowed into the Darwin and Panamint basins. These in turn were connected with Death Valley during periods of high water. And it is even possible that an old outlet from Death Valley emptied into the Colorado River near Parker, Arizona. The lake covering Death Valley at its height was several hundred feet deep. The evaporation of so many cubic miles of water resulted in the deposition of the salt and borax which lie deep in the alkaline playas of Searles, Panamint, and Death Valley. Tufa domes, built up by the growth of stonewort algae, now stand around some of these dry lakes above the highest levels of the present meager floodwaters. Such domes are still forming in Pyramid Lake and Walker Lake. Thousands of them were formed in the lakes of the Great Basin in Pleistocene time, and now show various stages of decay. In a mirage they become transformed into aery figures.

Coastal landscapes suffered great changes during the Pleistocene. Fault movements formed the Nevada and desert-basin ranges in the late Pliocene and early Pleistocene. Fault movements uplifted the transverse ranges in southern California and tilted the earlier rocks into great ridges and scarps (30). Sediments eroded from these new mountains accumulated rapidly to depths of fifty thousand feet in the narrow Ventura basin. The extraordinarily thick beds in the basin were laid down at two distinct intervals—all in the last million years or so (6).

So run the records of the Pleistocene: four chief stages of ice, with four milder intervals, of which the present is the fourth.

Pleistocene Parade

Below the icy mountains lie green vales watered by streams and livened by herds of game. Mammoths of huge bulk, elephants of old, tread ponderously along well-beaten trails, through willow thickets, brittle manzanita, buckthorn, and greasewood—pausing to pull down the branches of a leafy tree, to poke their trunks among the twigs and pluck the tender leaves. Mastodonts roam the pine woods, trunks aloft, seeking the soft new sprigs standing like candles on the boughs. Horses and bison graze the grasslands, pursued by lionhead jaguars and wolves. Barrel-bodied sloths, shaggy, unkempt, uncouth, browse beneath the oaks. One lifts a monstrous paw to clutch a branch and bring the foliage down, champing his jaws and peglike teeth upon coarse leaves and twigs. An agile cat form pounces on the sluggish beast. Her dagger teeth stab through the matted hair and thick, bone-studded hide. Blood spurts from wounds, deep gashed. The cat beast guards her prey while vultures circle down to perch and wait.

Snorting tapirs scrunch on padded feet along the river's edge, uprooting sedges with prehensile lips, and watching warily a lurking bear. A drove of peccaries, manes erect, tusks chattering, scuttle through the woods, scaring up turkeys that rattle their coarse feathers as they fly away.

A grumpy badger knocks at the doorway of a gopher's den. His excavating claws scrape and cascade the yellow soil. A garlic odor warns him of a rattlesnake within the hole. Quickly he backs away, shakes off the dust, and shuffles off to tear apart a wood rat's jackstraw nest, and nuzzle in the burrow of a squirrel.

A beaver slaps his tail upon the pond. Otters climb and slide the slippery bank. Ducks pilot their sly broods among the cattails. A herd of elk comes down to drink, sniffing the quiet air for warning scent of wolf or jaguar. Teal cross the evening sky on nervous wings. Squadrons of slowly flapping geese head southward, honking their calls in baritone. Quail rustle in the thicket. A bobcat sits in wait above fresh earthen mounds piled at the entrance of a hole, and eagles peer down, ready to dispute the spoils.

The hills are crisscrossed with the trails of fork-horned antelopes and slender camels pantalooned like llamas of the Andes. Here they browse and scamper. Southward across the desert go their trails—from lake to lake over the grass- and bushlands, where the giant armadillo, armored from head to tail, lumbers like a turtle in his shell, and where roam the yucca-eating sloth, the elephantine tortoise, and the prancing horse.

CHAPTER 9

Immigrants
and First Families

F ROM ASIA CAME THE DRIFTING HERDS, across the northern isthmus, thence through Alaska — Siberian emigrants: moose, caribou and bison (pl. 8), bighorn and mountain goat, pika, mammoth, and wolf. And following the herds came man. The musk ox drew southward to the plains when the ice crept over Canada.

From South America, early in the Pliocene, along causeways like Panama newly risen from the sea, came outlandish hosts: porcupines, ground sloths, giant glyptodonts, and armadillos. In California, drifters from the north came face to face with these drifters from the south. The faunas mingled. The first families — pronghorns, deer, horses, tapirs, camels, mastodonts, beavers, wolverines, badgers, skunks, bears, and coyotes — mixed with newcomers from far lands during the great influx of the Pleistocene. A few of the old-timers — the rhinos, the oreodonts, and the bear dogs — had disappeared before the onset of the ice.

Southward went the camels, horses, tapirs, deer, mastodonts, dogs, cats, weasels, and bears, to invade South America and exterminate some of the ancient life of that isolated land. And also from North America, camels, horses, and mastodonts migrated into Asia.

Migration across what are now the Alaskan straits took place in the interglacial as well as in the glacial stages. But the shallowest seas prevailed when the ice accumulation was greatest, and population pressures in Asia doubtless were highest at that time. The interior of Alaska and Siberia, free from ice, provided a corridor for the Asiatic exodus.

The life and landscape of California reached their climax in the Pleistocene. Then came a million years of invasion. A cavalcade of monster beasts sought out this land, occupied it for a time, then faded away or was rapidly extinguished by man.

Among the northern creatures were the bears, well adapted to survive the cold climate. Even these retreated before the ice that finally covered the center of the continent. An early California bear was a short-faced brute *(Tremarctotherium)* like the spectacled bear still living in the northern Andes. His remains were found in the caves of Shasta County, associated with oxen *(Euceratherium)* and with the mountain goat *(Oreamnos)*. Bones of a large, short-faced bear were scarce in the tar pits at Rancho La Brea, and remains of a huge one were found in the Irvington early Pleistocene.

The crag-loving bighorn, cousin to the argali of Central Asia, came in the Pleistocene and drifted as far as northern Mexico and Lower California. Bighorns are mountain dwellers and live on the heights in Alaska and western Canada, as well as in the Rockies, the Cascades, and the Sierra Nevada. The bighorn has been exterminated by the hunters on much of its former range in California. Small, shy bands still inhabit the desert mountains of southern California, the San Gabriel range, and the Sierra Nevada. No bighorn survive on Mount Shasta, and perhaps they never lived in the northern Coast Ranges.

Other immigrants were elephants of the mammoth type. Earlier, in the Miocene and Pliocene, the land had been occupied by great mastodonts, elephant-like animals with short legs, long heads, simple piglike teeth, and tusks in the lower as well as in the upper jaws. The mastodonts possibly survived until a few thousand years ago in some parts of North America. The mammoths undoubtedly reached North America by the Bering route and probably entered California from the north. Great herds wandered over the coastal lowlands and into southern California. They rarely entered the desert. Their remains are among the most abundant fossils of the California Pleistocene, and associated with them near San Mateo was a flora similar to that which prevails today at Monterey.

At some time, or times, in the Pleistocene, the Santa Barbara Channel Islands were connected with the mainland. This is proved by the presence there of elephant remains, which have been found chiefly on Santa Rosa and Santa Cruz islands. Most of these elephants were pygmies, scarcely larger than an ox. The bones and teeth so far discovered vary greatly in size and resemble the pygmy elephant bones from Celebes and the island of Malta. No pygmy elephants are known from the California mainland. It

PLATE 8. The giant bison (*Bison latifrons*) represents a species distributed across North America in the Middle Pleistocene.

therefore seems likely that dwarf mammoths on the islands were descendants of early, normal-sized immigrants that had been stranded there when the islands were cut off from the land. Dwarfing may have been due to overpopulation and resultant lack of food. The larger animals would have been the first to starve. Dwarf mutants more readily survived on scanty food. It is significant that the elephants have disappeared from those islands, though there are no large predatory animals there today—except man. This suggests that the factors which caused general extermination of the large animals of the Pleistocene must have also caused extermination on isolated islands as well as on the mainland. This would imply some extraordinary climatic changes or extermination by man rather than by disease or predators. A very dry period occurred about five thousand years ago, which few of the great Pleistocene mammals survived.

During changes of climate in California, there was generally a north-south movement of floras, scarcely ever an extensive east-west movement. Hardwood forests and tropical plants have not been found in the California Pleistocene. This indicates that here the range of climates was more moderate than elsewhere in North America. The failure of tropical, northern, and desert animals to reach the California coast is also evidence of a continued mild climate in the coastlands. Survival of desert types is also significant. With the spread of humid and of frigid climates, desert animals would find themselves eliminated unless they could rapidly change. Such types as the desert tortoise, the chuckwalla, the dune lizard *(Uma)*, and many desert plants must have found refuge somewhere during the Pleistocene—probably along the lower Colorado River and in Sonora.

The early Pleistocene land-vertebrate faunas of California are represented by an early stage (Irvingtonian) characterized by an absence of bison and the presence of types like the four-horned antelope *(Tetrameryx)* not known in the later (Rancholabrean) stage. The type fauna of the early stage, at Irvington, Alameda County, is being carefully collected and prepared by the enthusiastic Boy Paleontologists of Hayward under the able direction of Mr. Wesley Gordon (52c).

Long post-Pleistocene records have been preserved in some of the desert caves, notably Gypsum Cave in Nevada and Ventana Cave in Arizona. Gypsum Cave lies twenty miles east of Las Vegas, Nevada. It is a series of low galleries entering the side of a limestone hill. Although the cave was visited by tourists, for many years nothing unusual was noticed. This is strange, since the cave refuse contains immense deposits of ground-sloth dung, which lie on the surface in the rear alcoves. One of the ground-sloth skulls found in the cave lay almost in plain sight behind a fallen rock. Mingled with masses of dried dung were clusters of coarse yellow hair, claws, and bones of one of the smaller ground sloths *(Nothrotherium)*, and also bones of extinct hares, llama-like camels of two kinds, deer, bighorn sheep, wolves, condors, and smaller kinds of animals that inhabit the region today. Darts of prehistoric natives, and cooking places (hearths), indicate occasional but not prolonged visitation by natives before the sloths disappeared.

The oldest deposits are layers of "gravels, sands and silts washed in from the entrance by water at some time when the climate was very much more rainy than it is now . . . some of these deposits were more or less solidified and cemented by gypsum derived from the solution of the limestone by acid waters" (23). The later refuse accumulated to depths of more than ten feet when the cave was continually dry, as it is today. This accounts for the excellent preservation of the material, which comprises human artifacts of four or more successive cultures, along with the remains of plants and animals of species still living in the vicinity. Because of the presence of bighorn and other recent forms, no great age can be ascribed to the remains of sloths and other extinct animals in the cave.

Dart shafts and points, yucca-fibre string, hearths, burnt sticks, and cane "torches" with burnt ends — unquestioned evidence of human occupancy — have been found in and below the layers of sloth dung, together with remains of sloths, camels, and horses. Darts are of two kinds: an older form, made of elderwood, which does not grow in the region today, found in the sloth-dung layers; and a later type, made of desert arrowweed, found above the sloth layers in the more recent deposits near the entrance of the cave. Both types of dart shafts are decorated in the manner of the Basketmakers. But the oldest are not exactly like previously known Basketmaker types and belong to an older culture predating Basketmaker culture II.

The more recent deposits, laid down above the sloth layers, contain Basketmaker cultures II and III, like those found elsewhere in the Southwest. Above these lie painted Pueblo pottery and corn-grinding stones (metates) similar to those in near-by Moapa Valley pueblos. Finally, there are modern Paiute arrow points. The Paiutes say that they used to visit the cave to make spirit offerings. The modern Zunis also make pilgrimages to Arizona caves and throw painted ceremonial darts into them.

Analysis of the sloth dung indicates that the sloths had fed on the coarse, saw-toothed leaves of the tree yucca of the kind now growing twenty miles from the cave and about two thousand feet higher in altitude. The tree yucca is sensitive to temperature and is confined to a narrow zone in the higher deserts of the Southwest.

The evidence at Gypsum Cave, then, shows that ground sloths occurred on the desert and ate desert vegetation at a time when dart-throwing natives also lived there — certainly this was not earlier than the "close of the Pleistocene" — and that the climate at that time was cooler and moister on the Colorado Desert than it is today.

Finds of this sort demonstrate the survival of the great animals of the Ice Age in North and South America until a few thousand years ago. What caused the final extinction of the mastodont, the ground sloth, the camel, and the horse? Man himself may have had a hand in ringing down the curtain on the Age of Mammals in North America. Persistent hunting by man, of large, dangerous beasts such as mammoth and saber cat, might have been a factor. Grass and brush fires set by man do not seem a likely cause of their extinction (52a), for such fires are easily avoided by the larger animals in Africa today.

Black Death

On southern plains, below a crest of hills, lie sticky pools of tar, seeping and bubbling from deep in the ground, petroleum and gas, a glairy mass, black, sinister, to trap unwary birds and beasts. Beetles crawl at night across the margin of the pool. Their spiny feet sink in the oil; gas blisters pop and smear their bodies with crude stuff that plugs their breathing tubes. A hungry toad spies the black beetles floating on the pool and stalks the helpless prey. One foot sinks down, and when he pulls, then all his feet stick fast. He feebly struggles till his head is caught. The evil film creeps to his nostrils and his eyes.

A questing owl swoops softly in the dark to seize the toad and finds herself aflutter, her talons shackled in the tar. Slowly she sinks, her feathers all besmeared. A lean coyote trots on silent feet, trailing for rabbits at the black pool's edge. He sniffs the luckless owl, crouches and sneaks to seize the fluffy blob just out of reach among the beetles and the leaves. He tests the surface with extended paw, pulls back, and tries again. Hunger destroys his caution. Rashly he springs to grasp the owl, thinking to outwit the tar and reach the farther bank. But this maneuver ends disastrously. The tarry trap holds fast his paws. For him, exhausted and with lolling tongue, the struggle ends.

For centuries the pits engulf their prey. Mammoths splash in to cool their hides. They find beneath the water film a fluid stickier than mud, which clings to their broad feet and holds them fast. Their bloated bodies rest like islands in the pond, where vultures sit, and lazy carnivores too old or sick to seek a livelier prey. These too are trapped—wolves, saber cats, lion jaguars, buzzards, and greater creatures of the air.

Wild horses prance and camels stride across the plain. Bison of old, heavy-humped, light-flanked, are geared to wheel upon the wolf or snarling cat; but speed and cleverness are not enough in this dire place, where death attends each careless step and stumbling foot.

The silky black film shimmers in the sun. Gas bubbles rise and swell and pop like toy balloons. The heavy liquid circulates within death's cauldron like molasses slow, a frightful brew. The bones move with the mass and come to rest when years have passed. The spring of oil is choked, the seepage clogged, the liquid hardens into asphalt. Dust blows across it, and rivulets spread sand to hide the surface of the pool. The bones rest in their sticky tomb, bones of animals that lived long ago, the ancient populations of the southern plains now covered by the habitats of man where cars spin by upon an asphalt boulevard.

Pleistocene Cemetery

T HE ASPHALT PITS at Rancho La Brea, now on Wilshire Boulevard in the city of Los Angeles, are a cemetery of Pleistocene life. These pits are formed by springs of petroleum mixed with gas that bubbles to the surface from deep in the rocks. The fluid spreads out around the vents to form small ponds of tar where animals become trapped and their hard parts are preserved. Evaporation has hardened the tar in the older pits into funnels of asphalt sometimes as much as thirty-five feet in depth and packed full of bones from top to bottom.

Hundreds of thousands of bones of animals and many plant remains were entombed there (pl. 9). Among these, not one complete skeleton has ever been recovered. The bones have been separated and some have been crushed by movements in the viscous cauldrons. It is necessary to collect many bones of each kind of animal in order to get enough to assemble a reasonably accurate mounted skeleton—all mounted museum specimens represent several individuals.

Bones of carnivores, such as the dire wolf and the saber cat *(Smilodon)*, are unusually abundant in the pits. The animals were attracted to the carcasses of tar-trapped herbivores in great numbers and were caught themselves. Bones of vultures, owls, hawks, and eagles are common for the same reason. Small bird and mammal bones, beetles, lizards, turtles, toads, and bits of wood are there. Mastodont remains are scarce. Bones of Columbian mammoths have been found principally in one pit which had engulfed a whole "herd" (pl. 9). Bones of tapir and the short-faced bear are scarce.

PLATE 9. Scene at Rancho La Brea tar pits showing birds and mammals, some now extinct, whose remains have been found preserved in the asphalt. Giant vultures, *(Teratornis)*, Columbian mammoth *(Mammuthus imperator)*, saber cat *(Smilodon)*, a camel *(Camelops)*, owl *(Bubo)*, giant ground sloth *(Nothrotherium)*, and coyote *(Urocyon)*.

Horses, camels, prong-horned antelope, deer, bison, large and small ground sloths, coyotes, foxes, cougars, lynxes, huge jaguars often called "lions," skunks, weasels, rabbits, mice, gophers, and shrews have been obtained from the older, now inactive pits. The more recent pits contain the kinds of animals living today, including human bones supposed to have been Indian and associated with the giant extinct vulture, *Teratornis*. The surface pits, still active, have trapped insects, owls, ducks, vultures, ground squirrels, a circus elephant, and a small boy. Indians living at the site used to retrieve mired ducks from the surface of the pools.

The long record at Rancho La Brea seems to show a varying succession of climates in the latter part of the Pleistocene and continuing to the present. It is hardly possible that all the forms found in the tar pits could have lived at any one time in any one climate. Extreme desert conditions are not represented; there are no typically desert animals—no desert tortoises, and no desert lizards. Tropical conditions are also dubiously shown. The presence of tapirs and some of the raptorial birds may indicate at least one warmer period, but there are no crocodiles or other really tropical reptiles and birds. The range of climates was evidently not greater than the differences between Monterey and San Diego today.

Cooler climates seem to have been present, but reported "bishop" pines and "Monterey" cypresses may be misidentified. A single specimen of pileated woodpecker, and bones of other birds normally restricted to the northern forests, are significant. The combined evidence seems to point toward at least one period—perhaps a late one—of heavier rainfall, when the live-oak forest was more extensive and there were pines and mastodonts. More arid and warmer conditions are indicated by the juniper, hackberry, desert shrew, and peccary (44). One might think that during part of La Brea time—perhaps the earliest part—the climate was hot and dry, like that of the northern coast of Lower California today. Then came a warm, wet period in which tapirs lived, and then a cooler stage.

At all events, there is no need to assign great age to the older pits at Rancho La Brea. The succession of climates during the advance and retreat of the ice in the North from ten to twenty thousand years ago is what might be expected. Evidently, at least once in late glacial and again in post-glacial time, the climate of coastal southern California and the San Joaquin Valley was warmer and drier than it is now. This idea is supported by the isolated colonies of desert organisms in the San Joaquin Valley, and by the record at Rancho La Brea.

At Rancho La Brea, there were mammoths and mastodonts, herds of bison and brush-living birds, horses and camels, ground squirrels, coyotes, jack rabbits, cougars, ground sloths, saber cats and lion jaguars. Horses were the prey of the jaguars, ground sloths fell to the sabertooths, deer were stalked by the cougars, the jack rabbits by the coyotes, the bison by the wolves. These were the herbivore-carnivore groups. This was mammal life at the peak of its development.

Tar seeps also occur at Carpinteria on the Santa Barbara coast, where plant remains are more numerous because this was a forested region. Monterey and bishop pines grew near by, and their cones were washed into the tar pits. Along with them were birds of kinds now living in the northern forests — nuthatches, crossbills, Steller's jays, the Northwest crow, and the chickadee. Remains of open-field birds such as meadow larks and marsh blackbirds are uncommon at Carpinteria but are abundant at Rancho La Brea. The climate at the Carpinteria Forest was similar to that of Monterey today; possibly the time was a little earlier than the earliest Rancho La Brea pits (44), but it could have been later, in Tioga time.

Other large petroleum springs, still active, rise near McKittrick in western Kern County. Water birds are abundant, and extinct mammals similar to those at Rancho La Brea. Probably during the late Pleistocene there were marshes or playas like those at Los Banos today. The saltbush in the deposits is like the one growing today in this arid region. There is an ox, *Euceratherium,* not known from Rancho La Brea but found in the Sierran and Shastan caves and in Mexico.

Song of Man

What spirit endowed man with courage to oppose the ravening beasts? And when did man arise? Where were the paths, beset by stones and thorns, his bare feet trod?

In long-forgotten lairs, buried in crannies and in caves, hidden beneath the debris of ten thousand centuries, mingled with bones of saber cats, a skull is found. What was this thing with brain but slightly better than the ape, and teeth like those of men but heavier, to crush and chew raw meat and seeds? His head was balanced on an upright spine. He walked erect, arms swinging free, with grasping hands no longer used as feet, active in chase, ready to fight and kill.

Man ape, half-man, primeval father of the race, first of the meat hunters, first to leave the forest trees to hunt in packs, to chase the game and stalk the wild baboon, to strike the baboon with a club and bash the head, to pull the lean meat from the bones, to gnaw the stringy flesh and crack the bones, to lie at night within the baboon's cave, guarding his children from the saber cat.

On his slow journey through the Pleistocene, man fashioned tools and learned the use of fire. Stone tools at first were rudely made. A pebble roughened to a cutting edge would serve for digging bulbs from hardened earth, for hacking head and limbs from buck, or mashing marrow bones. In time, man manufactured pointed stones to puncture hides and kill slow-moving beasts. And then he fastened sharpened stones to shafts of wood, inventing spears to pierce the swifter game.

By flaking fine-grained stones, small tools of slender knifelike shapes to hold an edge were spalled from larger stones. These served as skinning knives and then for fleshing skins, when skins came to be used as clothes to fend against the thorns and cold. From well-trimmed flint cores useful axes could be struck; held first in the hand to chop coarse fibers, bark, or wood for building huts. Then handles could be split, and lashed fast to the blades with strips of fresh green hide, to dry and shrink upon the stone. *From implements like these arose the spade and hoe to serve the women working in the fields.*

Finally, in later times, the dog, domesticated from the wolf, stood watchful over camp and cave, and followed on the hunt, to smell out game and claim his portion of the kill. Beasts and fowls were tamed and bred for work, for food, for clothing and companionship.

Man moved from his old hunting grounds to till the loamy soil and plant the seed he long had harvested and now first sowed and bred to larger size and succulence for food. Roving flocks and herds he drove to pastures and enclosed in pens. Clustered huts formed cities. Boundaries of the family lands were fixed, and planting dates were settled by the movements of the sun and stars. Records were kept by scratching bits of bark or stone or slabs of clay. Rules and laws, sculpture and painting, music, dancing, and religion were fostered by a hierarchy of priests and elders, ministers, and petty kings living in palaces of mud and stone. Marauding tribes were held at bay by guards, forever armed, maintained by taxes in the city-states. Dominion spread by force of arms, and captive peoples cringed beneath the lash.

Fat lands, Egypt and Babylon, gave birth to cities—islands of comfort far safer than the savage world of stealthy beasts and wilder men. Fat lands begot fat, sluggish men; lean lands, lean men alert to meet the challenge of the wilderness. And from the hard, lean, northern lands came the first New World folk, twenty-five thousand years ago.

CHAPTER 11

Rise of Man

MAN'S STRUCTURE declares his ancestral life in trees. But the fossil record does not yet tell from what particular group of tree-climbing ape creatures man arose. He has the general characters developed by the primates in the course of their long monkey life; and he differs from other mammals in just those characters: loose-jointed limbs; grasping fingers with nails instead of claws or hoofs, a thumb opposed to the other fingers, friction whorls on fingers and palms; a poor sense of smell; good eyesight, eyes that look forward and form a single stereoscopic image, eyes housed in a solid socket of bone; the same number and arrangement of teeth as in Old World monkeys and great apes; and a well-developed, convoluted cortex of the brain.

Man differs from the ape folk in his erect posture, larger and more complex brain, small canine teeth, relatively short arms and long legs, curiously shortened toes on walking feet, a nonopposable large toe, lack of body hair, and continuously growing scalp hair. He still retains great skill as a climber and trapeze performer, the curiosity, love of play, and family life of the monkey and the ape; but he has become a talking, laughing, singing, mcat-eating, tool-inventing, fire-using, house-building creature, living in fighting groups, rapidly developing social life, and imposing on himself and others the curious restrictions of such life. And he conducts organized warfare—a thing unknown among his "lower" relatives.

The present fossils trace man back nearly four million years. Thirty thousand years ago the Cro-Magnons were much like primitive man today—

drawing pictures, sculpturing in clay and stone, clothing themselves in skins and beads, fashioning well-finished stone tools, carving in ivory and bone, following the game, and feasting on the horse, the reindeer, the cave bear, and the mammoth elephant. In stature and brain size the Cro-Magnons were slightly above modern man; but not in social life.

Two hundred thousand years ago the Neanderthals inhabited Europe and Asia, and had relatives in Java and Africa. They were a sturdy folk, crude, heavy-bodied, bull-necked, chinless, low-browed, with deep-set eyes and thick skulls housing a brain as large as modern man's but prolonged backward and low above the ears. Their teeth were massive, heavy-crowned, and set in powerful jaws. Their rough stone tools were unhafted. Innocent of the arts, they made no ornaments or paintings or stone scratchings that have been preserved. They were hunters and fishermen and doubtless also gathered fruits, seeds, bulbs, and roots.

Five or six hundred thousand years ago Peking man had already learned to use fire and chip the stones to make rude tools. He hunted ostriches and the big-jawed deer on the plains of China. He varied his diet with hackberries. His remains have been found in a deep cave, where he must have fended off hyenas and the dread saber cats that stalked at night. His skull was small, extended in the rear, and was slightly ridged on top as are the heads of some modern Chinese. His brain was about two-thirds the size of ours, and lay behind a sharply sloping forehead. The jaws and teeth were large for the size of the head, and were much larger in the men than in the women. Peking man's close relative, *Pithecanthropus,* lived contemporaneously in Java, where his remains are associated with primitive elephants, hippos, buffaloes, and large land tortoises.

Below the sand and gravel terraces containing Java man have been found the great teeth and jaws of an ogre that may have lived before the time of fire and the fashioning of tools. Also in China, similar remains have been found — of a King Kong man ape whose distant relatives ranged perhaps as far as southern Africa. In the caves of the Transvaal, massive man-ape jaws have recently been discovered. The big-toothed African man ape, *Australopithecus,* bludgeoned baboons and probably ate them raw — no fire or tools have been found with his remains. Perhaps next year, or fifty years from now, the ancestors of man apes will be found. Until then that chapter of man's history will be incomplete.

The Transvaal man apes must be predecessors of man, if not his actual ancestors. How will the earliest man be recognized when he is discovered? The man apes walked upright and evidently used clubs, but they are not known to have manufactured stone tools and are not positively known to have discovered the use of fire. They had brains scarcely larger than the largest apes, but their teeth were subhuman. By some authorities they are placed in the family of great apes, Pongidae; by others they are placed with man in the Hominidae. Do speech, upright posture, tool making, or the use of fire mark the beginnings of man? The spiritual qualities attributed to man can scarcely be inferred from the earlier fossils. Primitive social life is not confined to man, but the early human hunting groups may have developed enough social organization to make it impossible for less advanced social orders to compete successfully with them. For this reason it is likely that man from the beginning never had severe competition from any animals except his own kind. And it may be inferred that the man apes of Africa and Asia would have eliminated other potential ancestors of man except those closely related to themselves. Man arose through intersocial struggles among differing but related social groups. The groups that have been able to dominate, and to preserve their own internal structure by intense cooperation have survived.

Episode of Folsom Man

What wandering beasts and men have set strange feet in this new land? What trails have led here from afar—from Asia's steppes, across the straits, along the Aleut's chain, by northern wood and lakeland, endless miles through tundra, swamp, and snow, across the frozen rivers and solitary vales trembling to the trump of elephant, the surly growl of bear, and cough of timid bighorn?

What were the scenes along the first man-trail?

Pond of the beaver, aspen glade, thicket of willow, tangled deep, resound to thundering tread of hooves and clash of horns as bison rush to cover from the storm. Wolf howls echo in the wood, and stealthy footfalls softly press the powdered snow. The fevered pack, with lolling tongues and pointed ears erect, circle the stricken bull. Teeth chatter and jaws snap. The bellowing monster falls in a cloud of snow and a mound of leaping fur.

Hungry spearmen drive the wolves from the carcass of the bison, feast upon it, and move southward.

Through tawny plains the river flows, from glittering snow and ice on Rockies' crest and fern-hung gorge. At river's bend a troop of prong-horned antelope step warily to drink. In single rank they line the water's edge, their russet coats, that blend with leaf and grass, invisible but for a strip of white below each slender form. With quivering muzzles raised they turn and bound away, on limbs so slender that they seem to float in air. Upon their rumps the danger signals flash, like white chrysanthemums, to warn the flock that here coyotes prowl.

Spearmen, encamped along the southern Rockies, follow the herds.

From the wood a trumpet call shatters the quiet of the morning glade. As echoes die, the monster charges forth on swaying limbs, ears flaring

PLATE 10. Folsom Man appears in North America in small bands. A large extinct bison *(Bison taylori)* is hunted by a pair of dire wolves *(Canis dirus)* while a beaver *(Castor)* cowers behind a rock.

out, trunk held aloft, and curved tusks gleaming white. Huge bodies, pressed together in a mass of heaving heads and tusks and tossing trunks, crowd up behind to see who dares intrude upon the feeding ground. It is a bumbling band of piglike forms that stumble as they run, and, terrified, knock down their fellows as they tear away—unwitting culprits they, the peccaries. So with majestic tread on ponderous feet, soughing and sucking in the mud, the mammoths amble to the riverbank to drink and cool their bodies in the swamp.

Spearmen watch a troop of camels, old natives of the land.

Filing across the skyline of bald hills, a caravan—grotesque as from an oriental scroll—of gaunt and gawky camels in a row of rocking heads, long-necked and gaited slow; bewildered, melancholy, bored, resigned to feed on bramble and dry bush. Grunting complaints against the rocky way, the strange troop passes down the barren trail.

Wild horses prance on the plain.

Beneath the morning breeze the supple grass bows low and curtsies to an unseen hand. Shy of the gentlest touch it shrinks away, in rolling rhythmic waves, a landscape sea. In challenge to the wind, proud, prancing hooves pound grass to dust, and billowing manes and tails stream out like banners; bold heads on arching necks toss high; nostrils dilate and snort defiance to the sky—squadrons of prairie cavalry!

A lion jaguar stalks the herd.

Now comes a slowly moving form that slinks and creeps. Groveling low, and tense, it crouches, tail aquiver, ears laid back; its tawny coat the hue of sod and dust cloud of the flying herd. As if from a magic catapult, the body shoots aloft; sharp claws on spreading paws lash out to strike the velvet neck and withers of a colt too young to match the herd's impassioned pace.

This impact crumples trembling limbs, and squeals of terror bring the herd to halt. The fanged jaws are open for the kill, but guardian mother readies with her hooves and aims a kick. A circling gallop brings the stallion up. Enraged, he rears and spars with battling feet, so flashing swift that agile cat can scarcely dodge the blows. Snarling, the lion-head jaguar slinks away, to ease his hungry belly and bruised flesh at distant water hole and shady den where mewling kittens play among gnawed bones.

130

A condor wheels aloft.

The copper sun mounts high, and shimmering waves of air, heat light-
ened, rise as if reflected from a burnished sea. Across that sea and over
quivering butte and tree-clump swiftly glides a shadow shape. And there
below the sun, in circles, wheeling round, on never-flagging wings, a car-
rion vulture soars majestically. From some far mountain roost, nest on
high crag, or cliff tree perch, with pinions spread to catch the first breath of
morning breeze, the soaring bird sailed forth to rise on steady wings in
spirals to and fro.

The great bird discovers the carcass of a ground sloth.

What expert sight can spy the hidden carcass far below? Or is it scent of
rotting carcass, borne aloft on rising surges of the atmosphere, that breaks
this solemn flight on half-closed wings, in headlong, whistling swoop,
diving to target on the sward, leavings of feast of carnivore: thick hide,
bone-studded; shaggy, gray-green hair; bare ribs like barrel hoops, all
curved above the gore; the head awry; limbs worried and askew—all that
is left of monster herbivore?

A thunderstorm brews.

On mountain heights the thunderheads pile high, hushed and ominous;
and over distant plains, tall twirling shafts of smoky dust proclaim the
coming storm. Even the spiders in their fluffy down float gently to the earth,
and horned larks cease to sing. Foreshadowing the blast, the thunder
cracks and rolls, a raindrop dashes down, clouds darken, and the silent
black envelops land and sky. The lizard slinks to crannies deep, and prairie
dog and owl scurry to catacombs beneath the sod.

The hunters run to cover from the storm.

A wind whiff stings with sand and grit and cold. Cool drops splash dust
to mud, and mud again is melted down in sheets of rain. Across the hills
the tempest twangs its giant strings tuned to an elfin dance where ice balls
bounce and pirouette on rock and tree.

The hailstorm ends.

Quickly the streaming clouds roll by. The rainbow beckons, and earth
shines in favored spots of sun, sweet odors, fresh like new-turned soil,
enrapturing. Then comes a rush and roar as brown waves lash the thirsty

sands. Banks break and crumble to the dashing flood, and tree trunks toss to race the stones that grind beneath the waves.

A Folsom scout watches a doomed bison herd and summons his tribe.

Now toward the flood bank flows a roaring stream of hairy bodies, with thump of sodden hoofbeats and clack of interfering horns — the buffalo. Stampeded by the storm, in blind, headlong, panic-stricken flight, they cataract across the verge, stumbling to the flood below and churning sand to mud — mad bellowing herd, a heaving dam of crippled flesh.

Within the shelter of a rock the lookout stands, alert to spy the herd, and at his feet his precious spear, the symbol of his occupation and his clan. He shouts, and flings his arms aloft. The skin-clad hunters dash from shelter cave and copse and cooking fire. Spears carried high, they close upon the wretched herd. The tribe smells meat — choice broiling-ribs and fleece, all fat and juicy from the hump.

With skill of practice born, death-dealing spears strike in and out, through matted hair, to reach the heart or sever artery. Time after time the thrust deals home. Each point fixed fast upon the haft is free for further stroke — weapon of matchless artistry.

The carnage ends. The women with their baskets and flint knives select the choicest parts and leave the carcasses beside the lowering flood. Hides and tails stripped away, meat torn from ribs and loin, the picked bones settle in the mire, with here and there a spear point broken from the shaft — a prize to be preserved for future wonder and surmise.

Thus, Folsom man, hunter of elephant and buffalo, expert with flint and maker of the spear before the use of dart and bow left flaking of the flint for lesser skill, our pioneer of pioneers one hundred centuries ago, has left this evidence: his weapon and his kill.

CHAPTER 12

Before the Mayflower

M AN, AS A HUNTER, discovered, entered, and roamed in America twenty-five thousand years before Columbus, following herds that pressed into Alaska from Siberia. Among the first hunters was a skillful weaponmaker, known today as "Folsom man" and identified by the remarkable spear points found in association with skeletons of the animals he slew, now long extinct. Folsom points, cunningly fashioned, represent the highest known art of the expert stone flaker. The workmanship of these points, in which a broad longitudinal groove is flaked from each face, is unexcelled. It is unlikely that any of the early Pleistocene Old World men ever made such points. Nor have such accurately flaked implements ever been found in the Old World among the Neanderthal cultures of the Middle Pleistocene. The Aurignacian (Cro-Magnon) and Solutrean flints in their workmanship approach that of Folsom man, but none of the implements gives evidence of such consummate skill. Our early American, therefore, was a clever, intelligent type, comparable with the modern Indian and Eskimo.

Speculations about Folsom man are based on the following slender clues: one, his uniquely fashioned spear points; two, the distribution of these artifacts and their occurrence in deposits underlying the remains of later cultures; three, the association of the spear points with skeletons of extinct animals such as *Bison taylori,* the elephant, the camel, and the ground sloth.

Lacking evidence to the contrary, and judging from the widespread distribution of the points and consequently of the hunting techniques, there is reason to regard Folsom man as having lived in migratory bands. He constructed no known permanent towns or habitations. He probably raised no crops and had no grazing domestic animals. Folsom man far antedated the beginnings of agriculture in America (4a).

Near the village of Folsom on the Cimarron River in New Mexico, the Colorado Museum in 1926 uncovered skeletons of a herd of fossil bison buried in an old stream bed. Among these bones were found spear points of an unknown type. These are now called "Folsom points," and the culture they represent is the "Folsom culture."

Folsom points have been found at widely separated localities on the high plains. Cruder points with some of the Folsom characteristics have been picked up in Minnesota, Ohio, North Carolina, California, and Canada. Some of the finer Colorado specimens have been discovered with and within the skeletons of extinct animals, especially the moderately long-horned bison of the *Bison taylori* type. Some have been found with skeletons of camels, elephants, and sloths. It is abundantly evident that Folsom man lived on this continent in association with animals now extinct, and that he hunted these animals and used them for food. At Lindenmeier, Colorado, a number of points have been found at a camp site where these weapons and other stone tools were evidently manufactured (70). It is possible, but not likely, that the points were made by a few skilled workmen and were distributed widely among peoples who were not able to manufacture them. It is more probable that the distribution of the points represents the actual range or territory of the people who made them.

The Folsom point is a highest development of the technique of flint chipping. The ordinary Indian arrow points, and even the slender obsidian knives of the Aztecs, were much easier to make. Amateur arrow-point flakers can produce an ordinary point in a few minutes from any good piece of flint, obsidian, glass, or bit of crockery. The Folsom point, until recently, has not been duplicated by even the most expert flint workers.

Don Crabtree, a skilled artist in the manufacture of stone artifacts, spent three years trying to duplicate the Folsom point, before he succeeded in fashioning a perfect one. He says that some thirty operations must be accurately performed before the two large, final chips can be successfully removed from the flat faces of the point. The final flaking must be done under an expertly applied pressure of about seven hundred pounds. To achieve this pressure, the primitive artisan must have employed a lever and a strong vise.

The point was first roughly split from a flake taken from a lump or "core" of flint tempered by baking, presumably in the hot ashes of the campfire. The preliminary work, done by pressure in the hand as in the making of an arrow point, included the removal of a number of small chips from the edge. The scalloped, saw-toothed border thus created was then ground down to prevent breakage when the spear point was pinched in the vise. Then the base was carefully ground until it was flat enough to prevent the lever from slipping. Finally, the unfinished implement was placed edgewise in a vise, made perhaps from a cleft log or the crotch of a tree. The lever, presumably provided with a point of bone or antler, was then brought down exactly on the margin of the center of the butt. By long practice the point maker learned to guide and steady the lever, while applying great pressure, so that the flake would give way at exactly the right angle and in precisely the right direction. If the pressure were not accurately applied, the flake would either be too short or it would carry away the tip of the point. Removal of the second flake was even more difficult, for the blade had become weakened by the removal of the first large flake.

After the two long flakes had been split off, the butt was notched, and the point was ready to be hafted. The type of hafting is not known, but evidently all the skill and care in flaking was done with the purpose of preparing a more secure fastening for the haft or spear handle. It may be surmised that the handle was a pole of hard wood split at the tip, and that the split end was fitted over the two concave surfaces of the point and fastened there by gum or bitumen. Then, a collar or wrapping of fiber, sinew, or thin rawhide was presumably lashed round the butt to prevent the haft from spreading and slipping away.

Thus was formed a weapon which could have been used over and over again in dispatching large animals such as the bison. The hunters no doubt pursued the game on foot. From the nature of the Folsom deposit it may be inferred that the hunters surrounded the herds. Or they may have cut off individual animals and driven them over embankments so that the bison could be killed by spears after they had been disabled by the fall.

Folsom is one of the earliest types of man to be found on this continent. But crude stone tools from Sandia Cave in New Mexico lie in deposits below remains of Folsom. A skeleton of a girl, thought by her discoverers to be contemporary with Folsom, has been unearthed in Minnesota. The evidence is inconclusive, for in stature and appearance the people of Folsom time perhaps differed little from modern Indians. Some of the characters of the American Indians may indeed have been inherited from Folsom man.

The first men that entered America were presumably of a late physical and cultural type, because of the comparatively recent arrival, perhaps as long as twenty-five thousand years ago. They were of Asiatic ancestry and must have been hunters, since only hunting bands could have obtained food on the northern overland routes into America. These hunters evidently followed the great herds across Siberia and Alaska to the New World. They were the first human immigrants.

Before the advent of the thrown dart and the bow and arrow, the elaborate flint technique of Folsom passed into disuse and was never regained (39). With the invention of the thrown dart or atlatl, the hunter no longer had to stalk the game so closely. And, with the coming of bow and arrow, spear and dart throwing became obsolete, and still more effective hunting became possible.

Plant foods supplemented meat in the diet and became more important as the bands moved southward toward the tropics. Means of gathering and grinding seeds and other foods were improved. Fish and mollusks were eaten. Finally, with the increase of droughts and extermination of the larger marauding animals, such as the ground sloth, the elephant, the horse, and the camel, came the first widespread practice of agriculture, which changed the living habits of the people.

CHAPTER 13

Red Men
of the Southwest

BASKETMAKERS, throwers of the dart, appeared in the Southwest long after the Folsom man. Basketmaker debris lies above Folsom remains in some caves. The two types of people never intermingled. There must have been a procession of tribes in the long interval between the Sandia-Folsom hunters and the agricultural Basketmakers, but we know little of them. Some of these early ones have left scanty remains in California, on ancient desert lake beds such as Lake Mojave (13), and in stream gravels. Long, riffled points have been found in Sierran caves near where the famous Calaveras skull, of disputed geologic age, was discovered (8). Older California burials may date back four or five thousand years.

Pueblo cultures in the Southwest came directly after the pit dwellings of the Basketmakers; the two cultures were continuous and were closely related. With the first Hohokam and Pueblo culture came the bow and arrow, the introduction of stone dwellings, coiled pottery, cotton, and irrigation.

The southernmost Hohokam of Arizona seem to have borrowed a few customs and practices from their more advanced Toltec and Mayan neighbors—construction of stone dwellings, domestication of the turkey and the macaw, captivity of the eagle, growing of cotton and of corn (maize), also a game played with a bouncing rubber ball on a court. It may seem strange that the long-civilized peoples of Mexico and Central America had so little influence on the adjacent Pueblo and Hohokam, and much less on the primitive Indians of California.

The southern civilizations in Central and South America developed many native wild plants into highly modified, cultivated ones—the potato, the tobacco, the tomato, beans, and corn, to name only a few. They were deficient in the domestication of animals, though they had the dog, the llama, the alpaca, and the turkey. The breeding of highly developed cultivated plants demanded continuous attention over a long period of time. The slopes of the northern Andes and the tablelands of Central America were well adapted for this purpose, for many climates and differences of soil lay no farther from each other than a man could walk. And there were no large herds of herbivorous animals to raid the primitive crops.

The higher civilizations of Middle America, which, as we know from the Mayan calendar, began at least two thousand years ago, were doubtless preceded by primitive agricultural societies of long duration. Agriculture and pottery making commenced only two thousand years ago in the Southwest, probably as an introduction from the south, for they were presumably practiced two or three thousand years earlier in Middle America from Mexico southward into the Andes.

The immigration of man into California must have taken place several thousand years ago. The Mojave Desert during the latest deep filling of the now-dry lake beds seems to have supported a widespread native population associated with mammoths, horses, and camels, which, judging by the evidence at Lake Mojave and Pinto Basin (2), may have lived eight thousand years ago when rainfall was still sufficient to maintain the desert lakes. Perhaps the rock pictures on the desert were made at this early time when natives traveled and lived more easily than is possible in the scorched lands of today. Heavily petrified human skeletons apparently associated with the jaws of extinct camels have been found in the San Joaquin Valley, California, near Tranquility on the Kings River delta (29). And human bones have been found deeply buried in a Los Angeles storm-drain ditch in which mammoth remains have also been found, some distance away; it is not certain that these are of the same age.

The archaeology of California contains little readily datable material—there are few stylized pottery types and no logs in which tree rings can be counted. The age of some human material is still in doubt. The metate and the mano (seed-grinding stones) evidently preceded the well-fashioned mortars and pestles (acorn grinders). In early burials the body was usually extended and in the later ones it was flexed. Cremation developed rather late. The practice of fishing increased and hunting declined. Use of bone and then of shell tended to succeed the manufacture of stone artifacts. Abalone-shell beads, however, are not found in the latest burials, and nearly all the stone beads found are from the latest period. Flints fashioned by striking (percussion flaking) were much more prevalent in early times, and pressure flaking came in later. The use of charm stones and of quartz crystals was for the most part early. Stone pipes are found mostly in shell mounds of the late period.

Partly on these data, Professor R. F. Heizer has divided the central California culture period into three horizons: the Late, in three phrases, back to 1,500 years ago; the Middle, from 1,500 to 3,500 years ago; and the Early, from 3,500 to 4,500 years ago. These tentative and estimated dates are now largely confirmed by radioactive carbon (C^{14}) datings.

The early horizon is represented by extensive cemetery burials at Windmiller mound on the Cosumnes River, and by the material in the Topanga and San Dieguito sites (27b). Much remains to be discovered. Finders of burials or other Indian remains should leave the material undisturbed, to be examined by experts, while it still lies in the ground!

Man was evidently present in California or in Lower California at least 8,600 years ago. Sea shells (*Olivella*) found in caves of that age in western Nevada must have been obtained, by trade or otherwise, from the western seashore (27c).

Indian tribes bearing the bow and arrow entered California at least three or four thousand years ago—from northern forests, from the steppes and plateaus of Utah and Nevada, and from the deserts of the Southwest. Hunters and fishermen from the north woods remained in the forests and along the salmon streams. They separated into many groups and developed diverse languages and customs. Those from the east and southwest likewise became diverse, sedentary, and peace loving. Secluded among the many-valleyed hills, the tribes and tribelets, clustered in small villages, eventually became segregated into seven main language groups with more than ten times as many dialects. Such diversity in a small area scarcely exists elsewhere today, though it may have been commoner long ago when travel was more difficult and dangerous.

Customs likewise were extraordinarily varied. Three main culture areas are known to have existed: one on the Northwest Coast extending into British Columbia, a second across Utah and Nevada into central California, and a third across New Mexico and Arizona into southern California. But these larger areas were divided into many subgroups, each with its own peculiarities.

Almost no agriculture was practiced by the California Indians. Pottery was made only in the south. The northern groups depended principally on deer and salmon, caterpillars and earthworms; those of the central coast, upon shellfish, small game, various plants, seeds, and insects; those of the central valleys, upon acorns, seeds, manzanita berries, game, and grasshoppers; those of the deserts, upon the agave, lizards, tortoises, small birds and mammals, and the lake-dwelling larvae of flies; those of the Colorado River, upon fish, mesquite beans, and other plants of the river

PLATE 11. A hunter returns to a Chumash village on the Santa Barbara Channel. A young girl carries water in an asphalt-lined basket. Near the entrance to a dwelling, a woman is preparing a similar basket by rolling heated stones and asphalt in it. Behind her is a stone for grinding acorn meal, with a basket hopper; and beyond a man emerges from a sweat-house. Some nets and a baby hang from a live-oak tree. In the little cove, a couple is preparing to launch a canoe made of planks sewn together.

bottoms. Nearly all California Indians ground their food in stone mortars — acorns, seeds, even insects were so treated.

Shell mounds, numerous along the California seaboard and on the Channel Islands, were village middens (pl. 11). The first navigators — Cabrillo, Drake, and Cermeño — found the coastal tribes still living on these mounds. Some of the mounds contain remains of crockery and ironware: mementoes of early voyagers. Some are of enormous size; the sites must have been occupied for centuries. Some are interlayered by fluvial sediments. The bases of mounds along the shores of San Francisco Bay are now from three to eighteen feet below high tide; this, with other evidence, indicates an age of three thousand years or more. But the California mounds contain bones of animals common until recently, and only a few contain archaic artifacts. Modern types of people must have built up the mounds by gradual accumulation of kitchen refuse.

Mounds near the seashore are composed principally of shells of mussels and clams, mixed with bones, earth, and ashes. This shows that the coastal population gained much of its livelihood from the sea. No evidence of any great change in diet is found in the mounds, except that in the vicinity of San Francisco Bay the presence of a greater number of land-animal bones toward the top of the mounds may indicate that the supply of shellfish decreased as the human population became denser.

Some shell-mound artifacts suggest that trade was carried on a hundred miles away or more, others imply that certain ceremonies, possibly religious in nature, were performed. This evidence is fairly evenly distributed throughout the older and the later levels of the mounds. It may be inferred that the habits and distribution of the coastal California tribes have remained nearly static for at least three thousand years, disturbed only by the fatal incursion of the whites.

Few other continental areas in the world have enjoyed such salubrious freedom from destructive warfare and disturbing changes, for so long a period. And few people have maintained themselves with so little change. One may search in vain for much progress in the arts in this long period. And, if there have been any tendencies toward degeneration, these are not apparent from the evidence.

During their occupation of the land, the California Indians no doubt achieved most of the inventions and adaptations which now distinguish them from other tribes — the pitch-caulked plank boats which enabled the Chumash to reach the Channel Islands, the methods of leaching and preparing acorn meal, the use and manufacture of a variety of stone mortars, the perfection of coiled and ornamented basketry (35).

Once settled, with tribal boundaries established by tradition and rigidly observed, in an equable climate with a varied and plentiful wild-food supply, separated by deserts and forests from aggressive neighbors, and peaceable among themselves, the California Indians had few troubles or difficulties to stimulate them toward restless migration, invention, or warfare. The impact of the Spanish settlers created problems far greater than could be solved. The untutored people quickly succumbed and allowed themselves to be dispossessed. Today many of the ancient folk are extinct, others are on the brink of extinction, and the remnant of fifteen thousand is perhaps only a tenth of the number that lived in California two hundred years ago.

Native warfare took the simple form of feuds between families or villages. The Mohave Indians along the Colorado River engaged in more bloody combats, and sometimes made raids across the mountains. Ousting of one tribe by another was almost unknown, tribal grounds and boundaries were respected, and doubtless were maintained for centuries.

Wars and troubles with the whites ranged from full-scale attacks on troops, as in the Modoc War, and massacres such as the slaughter of Jedediah Smith's men at a crossing of the Colorado River, to individual strife and wholesale extermination campaigns by the whites, as in the Mill Creek "war."

Early white voyagers had minor difficulties with the coastal natives. The missionaries, with few exceptions, found conditions peaceable. The Spaniards stationed soldiers at the missions, however, and made "punitive" raids from San Jose, Sonoma, and elsewhere. The missions tended to separate the native population into two parts: subservient neophytes and semi-Christian natives attached to the establishments, and the wild pagan tribes that lived uncontrolled by white men. The labor of the natives built the missions, but the natives did not prosper under the new regime. The whites brought in diseases which almost exterminated the natives. Depletion of hunting grounds and disruption of food supplies caused widespread distress and starvation.

The Indian question is a live one. Most of the lands occupied by Indians are poor and restricted. Oil has not been discovered on the California reservations. Litigation over old treaties still goes on, and large payments have recently been granted. Payment is probably not salutary, and the need for land and educational facilities is paramount. The old Indian is gone, and the younger people find it difficult to adjust themselves to the modern scene.

CHAPTER 14

Invasion

W HITE MEN INVADED America by way of the West Indies and Mexico less than five hundred years ago. They conquered the natives by force of arms and fifth-column trickery. They soon destroyed the aboriginal civilizations of southern Mexico, Central America, and Peru. They exterminated the Inca and Aztec aristocrats and took over their lands and slaves. Seeking to expand their dominions, hungry for trade and greedy for gold, they slowly made their way overland toward the north and spread more rapidly by sea along the western shores.

Their ships, built on the west coast of Mexico, penetrated the Gulf of California in 1540 and sailed up the coast of Lower California nearly to San Diego. Two years later, Cabrillo, a Portuguese navigator in charge of two small Spanish vessels, entered San Diego Bay, landed on the Channel Islands, sailed along the coast as far as Monterey, and returned to San Miguel Island, where he met with an accident and died. His mate, Ferrelo, continued the voyage to beyond Cape Mendocino and then returned to Acapulco.

Cabrillo's expedition was a result of Cortes' conquest. Its purpose was to seek gold and treasure and to take possession of new lands for the Crown of Spain. Glittering treasures were not found. Instead of rich cities, there were only poor native villages, with nothing of value that could be carried in trade by the small ships of the day. Hence, Cabrillo's reports were not even published for many years, and his epochal voyage was all but forgotten.

Meanwhile, the Spanish and Portuguese had settled their differences regarding the division of the world. The terms of settlement prevented the Spanish from using the route to the Indies round the Cape of Good Hope. The Philippines, lying within Spanish domain, could best be reached from Mexico. The Philippine galleons took newly mined silver from Mexico to China and the Islands and purchased Chinese goods, and spices, to be sold at a good profit in Mexico and Spain. The long voyage was venturesome. Many ships were lost. The crews suffered miserably from scurvy. It was therefore decided to seek a port on the California coast that could be used for relief of the crews engaged in the transpacific trade. Gali and Unumuno had visited this coast. Cermeño, in a heavily laden craft from the Philippines, was wrecked at Point Reyes and made his hazardous way to Mexico in an open boat, shortly after the landing of Francis Drake. The Spaniards were worried about the activities of the English — their piratical attacks on shipping, their supposed access to the Pacific through the legendary Strait of Anian, and their claims to Pacific territory — all of which, doubtless, were reasons for Cermeño's visit. After the Cermeño disaster, it was thought best to send unburdened ships to explore the California coast, rather than to entrust that work to the tired crews of the heavy galleons returning from China and the Islands.

Vizcaíno was the first to map the coast. His expedition rediscovered and named Monterey Bay, but it by-passed the better harbor of San Francisco — which was not to be entered until more than a century and a half later, when the first Spanish settlements and missions were being founded in Upper California.

During this long interval a considerable trade in pearls had continued in the gulf, and missionary interest was active in Lower California, Sonora, and the Southwest. New Mexico and northern Mexico had been settled, and the way was clear for new contacts with California by land and sea. These contacts were due primarily to the zeal of the missionaries, especially the Franciscan, Junípero Serra, who became president of the Lower California missions upon their release by the Jesuits in 1768.

After the founding of San Diego and Monterey by Portolá and Serra, and the overland discovery of San Francisco Bay and settlement there, missions were rapidly established as far north as Sonoma. These flourished until the Mexicans won independence from Spain in 1821, when California became a distant province of the Mexican empire and finally a territory of the Mexican republic. Under the republic, the mission estates were delivered into secular control, and they rapidly disintegrated.

Meanwhile, a Russian fur company, seeking a favorable climate in which to raise crops and cattle to supply its bleak Alaskan posts, established an outpost near the mouth of the Russian River and there built Fort Ross. Russian sea-otter hunters cruised along the coast, bringing with them Aleut spearmen who hunted from bidarkas. These were the men who massacred the Indians on remote San Nicolas Island. The Spanish Californians, alarmed by the presence of the aggressive Russians, sought to strengthen their holdings in Sonoma County on their own northern frontier.

The weak California government also became alarmed over the activities of enterprising foreign traders, smugglers, and fur hunters who drifted in by sea and overland. Those who failed to show passports or other proper credentials were often clapped into jail. Hostile feelings led to trouble with Frémont's third expedition, and finally to the Bear Flag revolt and the conquest by American troops and settlers during the Mexican War in 1846.

The Treaty of Guadalupe Hidalgo, in 1848, ended the Mexican regime in California. Gold had been discovered at Sutter's mill only two weeks before the treaty was signed, but the signers of course knew nothing of this world-shaking event. Half-crazed gold seekers soon poured into California from everywhere. Mushroom towns sprang up—from San Francisco, across the valley and along the Sierran Mother Lode, and on a lesser scale in the Trinity region and in Shasta County (32c).

The new population, badly mixed, had difficulty in adjusting itself. Discriminatory laws and taxes, lynchings, Indian raids, vigilante movements, and finally labor agitation, created unrest. The old Spanish Californians were rapidly dispossessed. Shrewd operators exploited, preempted, and squatted. Fortunes were won and lost. The avid search for gold extended into Arizona, Nevada, Idaho, British Columbia, Colorado, and Alaska. And California miners were always to be found in the new diggings, even in faraway Australia and South Africa. Meanwhile, industrious farmers discovered the possibilities of the soil and climate—far greater resources than the gold.

The first governor of California was a Missouri farmer who had come overland, driving his own ox team. Most of the farmers came from the Middle West, particularly from Illinois and Missouri in the early days, then from Iowa, Kansas, and Nebraska, and in recent years from Texas, Oklahoma, and Arkansas—a steady inpouring for more than a hundred years. The early immigrants were attracted by natural resources which offered a tempting and liberal livelihood: the furs and game, the farm and cattle lands, the gold, the oil, the timber, and the climate. Later migrants have been stimulated by growing trade and industry, educational advantages, and old-age pensions.

The population has grown in spurts, by feverish immigrations. The land rush of the 'forties was followed by the gold rush of the 'fifties. Then came the railroad epic, the wheat boom, the southern real-estate booms, the oil boom, the motion-picture boom, the cotton boom, and the shipbuilding and airplane-building booms of wartime.

The population of California has rapidly risen to over twenty million. Populations have a way of increasing to the limits of the land and the industry available to support them. This has already occurred throughout most of the world. Nature, if left free to take her course, produces a China, an India, and a Puerto Rico. Population is commonly far in excess of the optimum for comfortable living. Countries with populations beyond the productive capacity of the land suffer disastrously if deprived of their industries. Those with but few industries and no birth restrictions are cut down at times of famine, war, pestilence, and flood. These calamities are the diseases and penalties of overpopulation, the fate of any overpopulated land where ingenuity and resourcefulness and birth control have not postponed the effects of crowding. California, still in her youth, can hardly envision so ominous a future. The country land is not yet surfeited. Life is pleasant for those who can afford comforts.

But some restrictions and critical shortages loom ahead, and one of these is the water supply. At this moment California is using nearly all available sources of water. A long-continued drought would be disastrous, despite all the engineering skill that has gone into the construction of dams, aqueducts, and canals. Engineers have considered a way of augmenting the supply: this is to condense the sea water, using cheap fuel directly at its source in the oil fields, or using atomic power. Salt, mineral, and biologic residues would help to pay the cost of such a project.

The rich, highly seasoned broth of the sea contains hundreds of dissolved substances. Each cubic mile holds more than five hundred million dollars worth of salts and metals. These include 14,000,000 tons of common salt, 3,500,000 tons of Epsom salts, 700,000 tons of calcium chloride, 364,000 tons of potassium, 294,000 tons of magnesium, 875 tons of iron, 840 tons of aluminum, 56 tons of copper, 30 tons of iodine, 8 tons of silver, 86 pounds of gold, and carloads of plankton and other material. The vast resources of the sea could support an increased population. But other shortages will eventually arise, and prudent policies of population and immigration control will be extended.

Man struggles against the environment and with his own kind to prevent his own extermination. He becomes careless and greedy and exploits the resources of nature. He wastes and destroys unwisely, altering the delicate balances in the life about him, seemingly heedless of consequences. He expends his patrimony and surfeits himself with pleasures. Man would like to control his destiny. He attempts this by defying and modifying his environment in countless ways not usual among other living things. He has set himself above the rest of life, and uses life for his own purposes. All this is courageous, if not presumptuous, of man. For Nature whispers to him saying, "Take heed or you shall perish." This warning is not easily heard; and if heard, it is not often heeded.

Violent overthrow of nature's checks and balances is an unforgivable sin. Such flouting of nature is to be seen in man's extreme alteration of the beneficent environment—the destruction and waste of resources, to gain temporary ease, pleasure, and advantage. There is also some prospect of decay of human vigor: the mental and physical effects of soft and crowded living, of lack of freedom for children to safely run and play, of the hectic excitements of city life, of the nerve-racking demands of organized business, industry, and politics, and of the reproduction of the obviously unfit.

CHAPTER 15

California on
the Changing Earth

C ALIFORNIA OCCUPIES a badly fractured and unstable segment of the earth crust, along a coast rising from the sea. Her coastwise landscapes, newly formed, are eroded under the impact of the combers pounding against the cliffs and headlands along her rugged shores. Drowned valleys have produced safe harbors at only two or three points. The breaking waves attack the land, undercut the cliffs, deposit the finer particles of rock and clay in deeper water, and leave masses of sand along the beach. The beach sand is continually reworked and transported by shore currents. These currents flow toward the south and drift in that direction.

A jetty was built at Santa Barbara to create an artificial harbor. After a few years the sand piled up in a broad new beach west of the jetty, and bathing resorts to the southeast became denuded and reduced to bare rock. After this the new harbor began to fill with sand that drifted round the end of the jetty. Thousands of tons accumulated, and the harbor was threatened, so that it was necessary to spend liberally for dredging.

The waves tend to trim the headlands, to fill the coves, and to straighten the coast line. They also cut the shore down to wave level at low tide; they have thus produced an offshore shelf which extends out a mile or so along much of the coast. Beyond is the continental platform, which continues out to the farthest islands and in most places is less than two thousand feet below sea level. Geologically this is a part of California; parts of it were once above the sea, and some of it at some distant day may rise again.

If the coast should be raised ten thousand feet, high ridges and deep canyons now beneath the sea would come into view. The largest of these canyons, off Monterey Bay, has precipitous walls like those of the Grand Canyon of the Colorado. It heads at the mouth of Salinas River and extends, six thousand feet deep, for many miles into the abyss of the ocean (fig. 29). Like other canyons along the inshore submarine slope. it extends almost at right angles to the shore line (57).

Echo soundings have made it possible to map more accurately the gorges beneath the sea. How these gorges were formed is not known; authorities disagree. Some think that they were made by submarine currents, by mud flows, and mud slides, or by excavation in soft material along submarine fault lines. Others believe that the canyons were caused by streams when the land, now submerged, stood several thousand feet higher. It is improbable that a system of canyons formed on land could have been preserved as open canyons while subjected to wave action during slow submergence. They would at least have been filled with sand and detritus while they sank, and much of their strong relief would have been worn away (56). But rapid sinking of an ancient Pliocene coast might account for the deeper canyons.

Possibly, the accumulations of waterlogged sand, mud, and ooze on the sloping sea floor, when sufficiently thick, tend to act like mud glaciers or mud rivers. They may gouge as they slip down the slopes of the continental platforms, and this might account for the clearing and deepening of canyons (16a). It seems probable that some nearshore cutting took place when there was much less water in the oceans than there is today, during accumulation of glacial ice in the Pleistocene. And excessive debris during the Pleistocene may have helped to gouge out the deeper canyons which are cut through Pliocene and older rocks (18).

Lifted above the foothills are great mountains, constantly wearing away under the expansion of ice, insidious decay, avalanches, and torrential streams. In these mountains are mineral veins long exploited for ores for which the miners dig deeper and deeper. In the hills are fossil fuels, oil and gas, which soon will be exhausted.

In the distant past, the water lay in great shallow bays along the California coast. Some of these bays covered what are now the lowlands of the south and the Great Valley. Some were evidently connected with the sea by narrow straits like the entrance to the Black Sea. In modern inland seas, swarms of tiny creatures sink to the bottom to form a muck which never completely decays, owing to lack of oxygen in the enclosed waters. The fatty and oily residues are least altered by bacteria and remain in the enclosing sediments to become petroleum, one of the most useful sources of energy for mankind. California lies in a region where quantities of petroleum have accumulated. Other similar deposits are found in unstable parts of the crust where sea waters have periodically inundated the land. One of these areas is along the margins of the Gulf of Mexico and the Caribbean, particularly in Louisiana and Texas, the east coast of Mexico, northern Colombia and Venezuela, and the island of Trinidad. Another oil center is the region around the Gulf of Persia, the Caspian Sea, and the Black Sea. The East Indies and Burma comprise another. And the shore of the Arctic is starting to produce oil.

Petroleum lies in the spaces between sand and particles of silt in the rocks. It flows about in the deep rocks under pressure of water and natural gas. It floats on the subterranean waters; it is consequently found in the summits of domes, or trapped in other closed structures. It is a fossil; the energy it now contains was originally produced by microscopic plants which used the energy of sunlight to transform simple molecules of water and carbon dioxide into the large, complex hydrocarbon molecules preserved in petroleum.

Petroleum has accumulated over a long stretch of time in California sediments, from ninety million down to fifteen million years ago. It is being used rapidly now, and the dwindling supplies cannot be replaced.

California is split, almost from stem to stern, by a break in the rocks: the San Andreas rift, an active fault zone that slices through Point Arena, Tomales Bay, Olema, and Bolinas Bay, passes offshore beyond the Golden Gate, bisects the Peninsula through Palo Alto and Santa Clara, traverses the Carrizo Plain, Tejon Pass, the north side of the San Gabriel Range, Swarthout Pass, the Cajon, the south front of San Bernardino Mountain, and San Gorgonio Pass, and finally reaches the Colorado Desert near Salton Sea and the Mud Volcanoes (fig. 29).

This break extends down some twenty miles through the crust. Southern California is slowly drifting toward the sea, in a northwesterly direction with respect to the main body of land to the northeast of the rift. The rate of movement is estimated at about twenty feet a century, and the elasticity of the rocks permits the movement to proceed and to build up tremendous stresses along the fault. Friction along the deep crack prevents slipping, except at long intervals when accumulated pressure causes the fault to slip, creating an explosive major shock, which is usually followed by minor, intermittent aftershocks. The sudden release of elastic energy throws the crust into convulsive waves; these lose force as they travel outward, and usually become harmless in less than thirty miles.

Viewed from the air, the San Andreas fault looks like a straight furrow ploughed through the landscape — up over mountains, across the plains, dipping into the sea. From the ground it is a strange mass of tumbled hillocks formed of slivers of rocks and rock fragments which, blocking watercourses, form lakes and ponds and give rise to springs. Movements

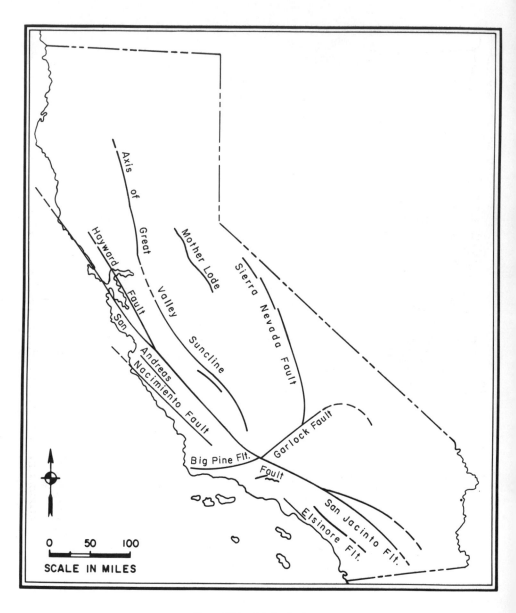

FIG. 29A. Main faults.

150   CALIFORNIA ON THE CHANGING EARTH

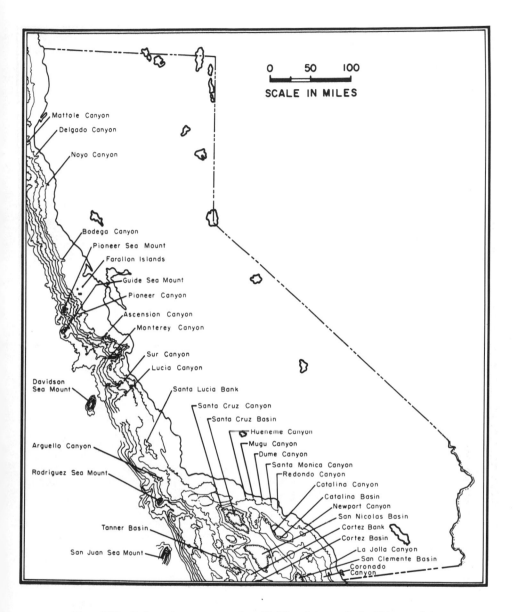

Scale: 0 — 50 — 100 SCALE IN MILES

Mattole Canyon
Delgado Canyon
Noyo Canyon
Bodega Canyon
Pioneer Sea Mount
Farallon Islands
Guide Sea Mount
Pioneer Canyon
Ascension Canyon
Monterey Canyon
Sur Canyon
Lucia Canyon
Davidson Sea Mount
Santa Lucia Bank
Santa Cruz Canyon
Santa Cruz Basin
Hueneme Canyon
Arguello Canyon
Mugu Canyon
Dume Canyon
Rodriguez Sea Mount
Santa Monica Canyon
Redondo Canyon
Catalina Canyon
Catalina Basin
Newport Canyon
San Nicolas Basin
Tanner Basin
Cortez Bank
Cortez Basin
La Jolla Canyon
San Juan Sea Mount
San Clemente Basin
Coronado Canyon

FIG. 29B. Submarine topography. California's eroded offshore platform is geologically a part of the continental margin; parts of it have
stood above the sea for long periods. It is intersected by submarine
gorges and contains deep depressions and high mountains. (From
Shephard and Emery, 57.)

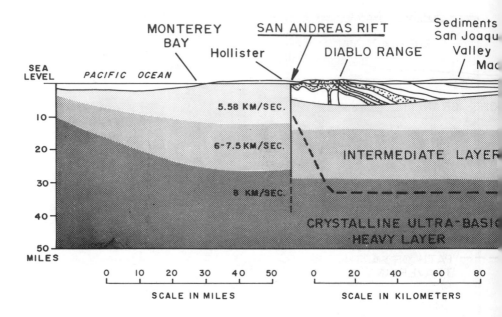

MONTEREY BAY · SAN ANDREAS RIFT · Sediments San Joaqu··· Valley

Hollister · DIABLO RANGE · Mac···

SEA LEVEL · PACIFIC OCEAN

5.58 KM/SEC.

6-7.5 KM/SEC.

INTERMEDIATE LAYER

8 KM/SEC.

CRYSTALLINE ULTRA-BASIC HEAVY LAYER

10 — 20 — 30 — 40 — 50 MILES

| SCALE IN MILES | 0 10 20 30 40 50 |
| SCALE IN KILOMETERS | 0 20 40 60 80 |

FIG. 30. Cross section through central California from Monterey Bay to Owens Valley. Thick, contorted sediments form the Diablo Range; thinner sediments are still being laid down in the San Joaquin Valley; the granitic basement thins toward the sea and thickens under the

occur at different times in various segments of the rift. The San Bernardino earthquake of 1857 was localized in the south. The great earthquake of 1906 was a rupture of the northern part of the fault, from Navarro south to San Benito County. At Olema in Marin County, near the southern tip of Tomales Bay, the maximum horizontal displacement, measured by an offset road, was about twenty-one feet. A corral stood over the fault line, where the earth opened in a wide crack and engulfed a cow. Milkers at work found themselves rolling on the ground. A row of trees standing by the farm gate marched past the farmer's porch. Earth cracks broke the roads apart and made them impassable. Near Fort Ross a large redwood tree was split in two like a wishbone. A lumber town shabbily built on a soft delta at the mouth of the Navarro River was knocked about like a set of toy building blocks.

Paralleling the main San Andreas fault is a broad belt containing other smaller faults. The San Jacinto and Elsinore faults south of Riverside, the Nacimento fault south of Monterey, and the Hayward fault along the west

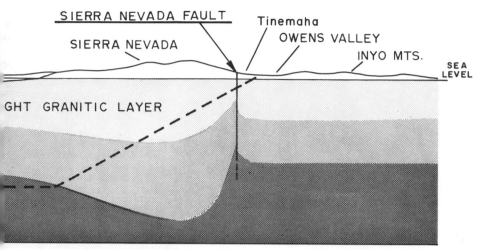

SIERRA NEVADA FAULT

SIERRA NEVADA

Tinemaha

OWENS VALLEY

INYO MTS.

SEA LEVEL

GHT GRANITIC LAYER

- - - PATH OF EARTHQUAKE WAVES, DELAYED 4+SECONDS IN
TRAVERSING THE "ROOT" OF THE SIERRA NEVADA.

Sierra to form a "root" upon which the Sierra floats high. Earthquake waves are slowed down when traversing the "root" and indicate its thickness. (Data from Bylerly, 11; Gutenberg, 22; and Taliaferro, 63a.)

base of the Berkeley Hills seem to be parts of this fault system (32a). Still smaller faults interlace and branch in all directions; many of them are inactive. South of the Tehachapi, the Garlock fault extends into the desert past Randsburg (fig. 29). Another great fault runs along the east foot of the Sierra Nevada, and recent vertical movements on it seem to show that the crest of the Sierra is still rising. Earthquake waves passing from the San Andreas fault to Owens Valley are slowed down by the lighter granitic rocks under the Sierra, and these rocks extend down to form a root or raft which lifts the mountains upward. For the mountain with its deep root floats like an iceberg in the heavier mass (fig. 30).

Adequately reinforced buildings along fault zones have survived the worst shocks. Building codes and inspections are now designed to minimize losses of life and property; but safety in the next quake, which may come at any time, requires vigilant attention to the condition of buildings, especially the large structures where crowds congregate.

Volcanoes are safety valves for the explosive materials of the earth's

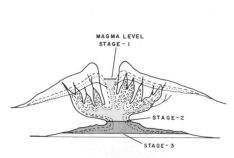

MAGMA LEVEL
STAGE - I

STAGE - 2

STAGE - 3

STAGE 1, HIGH MAGMA LEVEL IN VOLCANO NECK,
AND INNER CAVITY FILLED.

STAGE 2 AS MORE VIOLENT EXPLOSIONS OCCUR,
THE INNER CAVITY IS EXPANDED, MORE
MAGMA IS EXPELLED. LEAVING THE MAGMA
LEVEL IN THE MAIN CHAMBER EVEN LOWER

STAGE 3 ONCE THE INNER CHAMBER HAS BEEN
EMPTIED OF MAGMA, THE ROOF IS THIN
LEFT UNSUPPORTED.

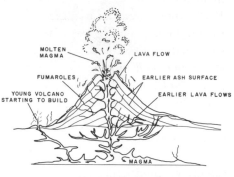

MOLTEN MAGMA

LAVA FLOW

FUMAROLES

EARLIER ASH SURFACE

YOUNG VOLCANO
STARTING TO BUILD

EARLIER LAVA FLOWS

MAGMA

VOLCANIC ERUPTIONS RANGE FROM, GENTLE
OUTPOURINGS, TO VIOLENT EXPLOSIONS THAT
MAY THROW ASH MANY MILES INTO THE AIR.
THE EMERGING MAGMA MAY BUILD UP A
MOUNTAIN OR FLOW SMOOTHLY TO FORM A
LAVA PLATEAU.
OUTPOURINGS OF MUD, ASHES AND GAS ARE
ASSOCIATED WITH VOLCANIC ACTIVITY.

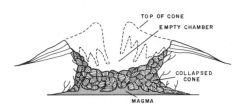

TOP OF CONE
EMPTY CHAMBER

COLLAPSED
CONE

MAGMA

STAGE 4 WITHOUT THE SUPPORT OF THE MAGMA THE
TOP OF THE VOLCANO COLLAPSES INTO THE
EMPTY CHAMBER. THEREBY FORMING A
CRATER. ONE OF THE BEST KNOWN EXAMPLES
IS CRATER LAKE IN OREGON.

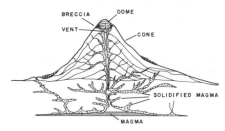

BRECCIA
DOME
VENT
CONE

SOLIDIFIED MAGMA

MAGMA

LAVA WHICH IS TOO VISCOUS TO FLOW ACCUMU-
LATES AROUND THE VENT TO FORM A STEEP-
SIDED PLUG CALLED A DOME. AS THE DOME
GROWS THE OUTER CHILLED SURFACE IS
FRACTURED AND FRAGMENTS TUMBLE DOWN
TO FORM A HEAP OF BRECCIA AT ITS FOOT.

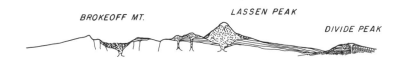

BROKEOFF MT.

LASSEN PEAK

DIVIDE PEAK

FIG. 31. Two types of volcanoes in the Lassen region, and their inferred history.

interior. Ancient volcanoes are numerous in California, but nearly all of them are dead or inactive. The volcanic explosions at Lassen Peak in 1915 represented the dying phases of activity in the southern Cascades. Lassen Peak is plugged by a dome-shaped core of solidified lava which was squeezed up like tooth paste into the old crater. Elsewhere in the region are similar lava domes (68*a, b*). Brokeoff Mountain, near by, is another type of volcano, a collapsed cone or caldera similar to Crater Lake (fig. 31).

Until six thousand years ago, the mountain which now holds Crater Lake in southern Oregon was much higher, and its flanks bore glaciers that carved polished troughs, the lower slopes of which can still be seen on the outer rim of Crater Lake. The great ancestral volcano, Mount Mazama, rapidly erupted material from beneath its summit, leaving a hollow space which could not be filled before the top of the mountain caved into the hollow throat of the volcano. This tremendous collapse formed the circular bowl where Crater Lake now lies (68*c*).

Man's manifold activities have altered the surface of our land. The Wilmington harbor district attracts attention because so much petroleum has been pumped out from below that it now is sinking, endangering the port facilities. Rivers and bays have been filled with sludge from the gold sluicing. Man-made fires and overgrazing have increased erosion on mountain slopes. Man transports and lays down sediments such as road pavements and refuse dumps. He pollutes the air and the streams and the sea. By draining lakes and subsurface waters he creates new deserts; he transforms other deserts into irrigated fields. He excavates canals, tunnels, and road cuts; dredges new harbors; blows up one island and builds up another near by. He overcultivates the arid lands and sees them swirl away in dust.

That is the geological position of California on this changing earth.

CHAPTER 16

California's Changing Life

C ALIFORNIA STANDS ALONE, cut off from the rest of the continent by wide deserts, high mountains, and the northern forests. Her climate is moderated by the sea and by the fringing wall of the Sierra. She lies between the humid tropics and the snows of the north. She has no frigid arctic weather except on the highest mountains. Great swamps, tundras, wind-swept steppes, and tornadoes are not on her list of attractions. Wild life was once rich and varied in this paradise.

Not so long ago her hills and valleys teemed with game, her waters with fish and fowl. Her pines and redwoods were the admiration of the world. Great schools of valuable fur bearers — seals and sea otters — and hordes of equally valuable whales once swam off California's shores. Now the salmon are gone from many streams, the beavers are much restricted, the grizzly bears are gone, the gray whales have diminished in number and the Guadalupe fur seals have been killed off. Condors soar infrequently above the southern ranges — their last outpost. Sea otters, thought to have been exterminated, have reappeared, and their furs could again become a valuable crop. The elk, the antelope, and the bighorn have been extremely difficult to protect and now exist only in small isolated bands, having been driven from much of the range over which they once ran in hundreds.

Hoof-and-mouth disease has spread from domestic to wild animals, rabies from pet dogs to coyotes, bubonic plague from ship rats to wood rats and ground squirrels. Other diseases lurk and tend to spread. Tick fever, tularemia carried by wild rabbits, San Joaquin fever—a lung fungus—and flukes and other parasites introduced by travelers from the tropical world are here to plague us. Money and effort are spent to prevent the spread of new pests and diseases. Air transportation requires a redoubling of precautions, especially against flying insects that may lodge in airplanes: Japanese beetle, boll weevil, alfalfa weevil, gypsy moth, and a host of others. What a price has been paid for pine blister rust, a deforester that runs a race of destruction with sawmills and fire!

Summers in California are characteristically dry. Except in small areas along the north coast and in the mountains, there is little or no rain between the end of May and the first of September. Since grass dries up in May and June, forest, brush, and grass fires become prevalent in the summer and early autumn.

Charcoal lumps in Pliocene sediments show that forest and brush fires occurred more than eight million years ago. Sporadic fires have evidently had a peculiar effect upon the flora. Plants which survive today are not liable to be exterminated by fire. Mature seeds of the annuals are as a rule unharmed by the rapidly moving grass fires. Native lilies in great variety grow from deep bulbs. The native bushes of the elfin forests—chaparral, ceanothus, manzanita, and other shrubs of the southern Coast Ranges—have chunky storage roots which head up at the surface of the soil and vigorously sprout again after the stems have been burned away. Native oaks and pines in the fire belt have thickened bark—that of the oaks is juicy—as a protection against fire damage, though the trees are not immune. The knobcone pine pops its seeds only when scorched by fire: the cones then explode to reseed the newly burned landscape. The thick fibrous bark of the redwood trees is fire deterrent. Redwoods survive fires that burn out the heart of the tree. The stumps of the coast redwood continue to sprout through thick layers of charcoal, and young sprouts will encircle the base of a burned stump.

Fires are infrequent in the humid north-coast forests and on the open deserts where grass is sparse and the plants stand far apart. But fires are destructive in the pine forests. Most desert plants are easily killed by fire, but the fan palm will resprout at the crown after a hot fire has destroyed its apron of dead leaves. Fan palms, badly burned, survived the Berkeley fire of 1923. Some of these stood near asphalt paving and underground wooden culverts that were consumed in the terrific heat.

These fire-resistive plants must have taken time to develop; probably the summer climates were dry in the late Tertiary. The plant fossils from earlier periods show that some of the forest types found in only a few places today were once widely distributed. These include the coast redwood, the tan oak, and the madrone. The redwood and its relatives were almost world-wide in the Miocene, and have been found as far north as St. Lawrence Island in the Bering Sea. The preservation of these trees—now extinct except in China, California, and extreme southern Oregon—is probably due to the mild, foggy, coastal climate. The giant sequoia of the Sierra has been preserved in moist spots which were free from summer ice in the late Pleistocene.

However, a few types, such as the fig and the magnolia, now living in other regions, were present here in Miocene and pre-Miocene times. Miocene climates were therefore warmer, moister, and more equable than those of today. Forests extended into northern Nevada and eastern Oregon. Much of our present desert was then a watered steppe, with horses, rhinos, antelopes, and camels in abundance—possibly like the plateaus of East Africa. Typical deserts probably occurred farther south in Mexico, for the surviving desert animals and plants are evidently old types.

Before Miocene times, in the late Eocene and the Oligocene, the climate was similar in parts of southern California—from Death Valley to the coast—to that on what are now the higher plains of Wyoming and Montana. The forests were lush and tropical, with large rivers, swamps, and lakes. The climate was moist and hot. Lemur-like animals similar to the tarsioids of the tropics today lived in what is now Ventura County. Fossils of giant, browsing titanotheres resembling rhinos have been found in Titus Canyon near Death Valley. These are also found abundantly in the Oligocene badlands of the Great Plains. Remains of primeval carnivores antedating the present cats and dogs have been found in Oligocene beds of Los Angeles and Ventura counties. These are also similar to those found in beds of the same age in Wyoming and Colorado.

Stream deposits (red beds) in which these bones occur seem to have been laid down in a warm climate; possibly there was a rain forest on the adjacent highlands and the tropical animals lived there rather than on the open plains below. The Pacific coasts of southern Mexico and Central America have similar combinations of dry and moist tropical conditions today.

California's coniferous forests with their glorious redwoods, sugar pines, ponderosa pines, spruces, firs, hemlocks, and a wonderful variety of other trees are among the world's finest. They are remnants of much greater forests that once extended across the West and into Asia. A narrow belt along the seaboard preserves tiny stands of strange pines and cypresses that may have had wider ranges in the Pliocene. Professor H. L. Mason believes that these trees originated in lands now under the sea along the continental platform.

The California timberlands are not only being depleted but are evidently decreasing in extent. The trees are delicately adjusted to the local climates and, by the shade cast over their roots, tend slightly to modify these climates. When marginal forests are slashed down, brush and grass may replace them; these hinder reforestation. Further cutting without replanting should not be continued.

Serious also is the depletion of the grazing ranges. Grass and delicate forage plants which once grew on many slopes and mountain meadows no longer exist. Most of the original forage has been replaced by more hardy foreign plants—mustard, wild oats, alfilaria, thistles of many kinds, and a host of other beneficial and noxious foreign weeds.

Thus, under man the living environment has greatly changed. This is the biological position of California. She is awake to the danger of depletion of natural resources and is trying to prevent wastage and spoliation.

Destiny

California! Four brief centuries ago you were a distant Siren on an unknown coast, tempting distressed galleons on the long route to Spain's Indies.

Two hundred years later you supported lonely outposts on Christendom's remote frontier, where leather-jerkined Spanish cavaliers met fur-capped Russians from Alaska. On your fair form the girdle of white settlement was first laced round the globe.

The Russians sailed back to their fur and fishing grounds. The Spanish Californians, subdued by Yankee troopers in brief skirmishes, lost their feeble hold upon your coast, and Pío Pico fled to Mexico. England peacefully partitioned her Northwest domain. Smoke of great battles never dimmed your skies.

A hundred years ago, James Marshall picked a flake of gold from Sutter's millrace at Coloma. A fevered world sent eager young adventurers by land and sea to seek their fortunes on your Mother Lode. And Statehood's crown was placed upon your golden head.

Bastioned by mountains, shut off by wilderness, for twenty years you dwelt alone. Then from the tumult of those adolescent years came your marriage to the Union, when shining rails strung on canyon cliffs and across the wide desert joined you to the East. Infant midwestern settlements grew up in the shadow of your strength.

T.R. built the Panama Canal. The shortened sea lanes, the highways, and the airways bound you even closer to the heartland of the East.

All this occurred in time to stem the Oriental stream of conquest. Asia's tide was turned by Europe's westward flow — across America, into the Pacific. Old Asia's tide surges again, as it has surged before. Now, while other sirens scream across your rooftops, you play your part in destiny.

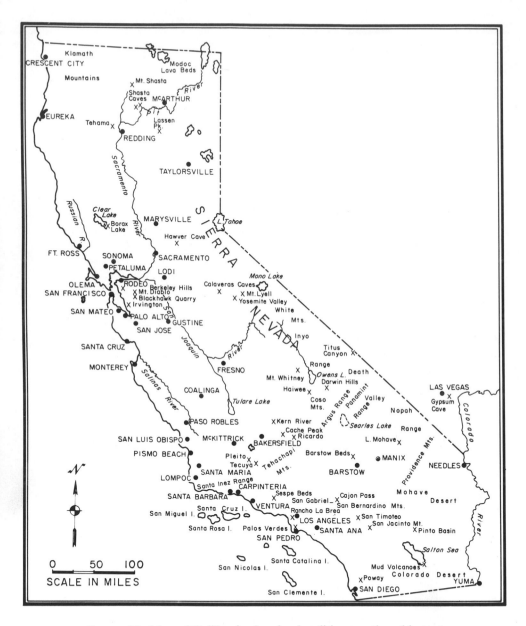

PLATE 12. Map of California showing localities mentioned in text.

162

Charts

CHART 1. Geological time has been subdivided into eras, periods and epochs. The scale shows the time relationships of the main events of fossil history. The duration of each period and the elapsed time, in years, since the beginning of each large subdivision are given in the left-hand columns. Most of the figures are rough estimates. The only satisfactory determinations based on the radium clock are: the Upper Cambrian — 460,000,000 years; the Lower Permian — 230,000,000 years; and the late Paleocene — 60,000,000 years (34). Other dates are based on less accurate readings of the radium clock, and on sedimentary and fossil records (71).

Nevertheless, the sequence is well established. The chart mentions some of the stepping stones in the history of life: the appearance of primitive life in the Algonkian; the rapid expansion of hard-shelled invertebrates in the Cambrian; the advent of backboned animals in the Ordovician; the development of fishes with jaws in the Silurian; and emergence of amphibians and plants on land in the Devonian; the abundance of insects, reptiles, and coal forests in the Carboniferous; the rise of reptiles and primitive pines in the Permian; the first mammals, dinosaurs, ichthyosaurs, and ammonites and the spread of monkey puzzle (auracarian) pine forests in the Triassic; the first known birds, and the rise of cycad "palms" and ginkgo (maidenhair) trees in the Jurassic; the disappearance of dinosaurs and large sea reptiles and ammonites in the Cretaceous, together with the rise of diatoms and flowering plants in that period; the rapid evolution of placental mammals in the Paleocene; the appearance of modern families of mammals in the Eocene; the evolution of plains grasses and long-toothed, grass-eating mammals on the grassy plains of the Miocene; the development of heavier, single-toed horses in the Pliocene; the coming of man with his stone tools and his use of fire in the Pleistocene; and finally, the domestication of animals and cultivation of plants, and vast improvements in tools and other devices, in the Recent epoch.

CHART 1.* Geological Time Scale

Millions of years (mostly estimates)		Era	Period	Epoch	Animals	North American plants and climates
Duration	Elapsed time					
			Quaternary	Recent	Domestic animals	Cultivated plants
2	2	Cenozoic Age of Mammals	Quaternary	Pleistocene	Man; extinction of large mammals	Deserts and tundras Glaciers in the north
10	12	Cenozoic Age of Mammals	Tertiary	Pliocene	First ground sloths and one-toed horses in North America	Cooler climates in the north
13	25	Cenozoic Age of Mammals	Tertiary	Miocene	First long-toothed horses	Plains grasses
15	40	Cenozoic Age of Mammals	Tertiary	Oligocene	Titanotheres Monkeys and apes	Temperate forests
20	60	Cenozoic Age of Mammals	Tertiary	Eocene	First horses, camels	Lush rain forests; warm and subtropical
5	65	Cenozoic Age of Mammals	Tertiary	Paleocene	Archaic placental mammals	Palms, figs, magnolias; tropical
70	135	Mesozoic	Cretaceous	Age of Reptiles	Extinction of large reptiles	Diatoms, flowering plants
45	180	Mesozoic	Jurassic	Age of Reptiles	First birds Flying reptiles and crocodiles	Cycads and ginkgos
45	225	Mesozoic	Triassic	Age of Reptiles	First mammals Ammonites Dinosaurs	Araucarian forests
45	270	Paleozoic	Permian		Early reptiles	Pines (*Walchia*)
80	350	Paleozoic Carboniferous	Pennsylvanian / Mississippian	Age of Amphibians and Coal	First reptiles Stegocephalian amphibians	Lycopods and ferns Coal forests and swamps
50	400	Paleozoic	Devonian	First amphibians Age of Fishes	First insects Lungfishes and crossopt fishes	First land plants
40	440	Paleozoic	Silurian	First jawed fishes	Arthrodires Acanthodians	Seaweeds
60	500	Paleozoic	Ordovician	First vertebrates	Ostracoderms Graptolites Corals	Abundant lime-forming algae
100	600	Paleozoic	Cambrian	Age of Invertebrates	Trilobites Sea snails Cystoids	Stoneworts
	+1000	Pre-Cambrian	Algonkian	Fossils rarely found	Sponges Tube worms Radiolarians Jellyfishes	Algae Bacteria
	+4000	Pre-Cambrian	Archean	Organic carbon in small oval masses		

*J. Laurence Kulp, "Geologic Time Scale." *Science*, vol. 133, no. 3459, April 14, 1961, 1105–1114.

CHART 2. Geological Events

Era	Period and epoch		California	Central and eastern North America	Mountain building
Cenozoic	Quaternary	Recent		Retreat of Sierran glaciers. Drying of desert lakes	Retreat of ice sheet	Rejuvenation of Sierra Nevada; Pasadenan and Coast Range orogeny
		Wisconsin — Tioga; Illinoian — Tahoe; Kansan — Sherman; Nebraskan — McGee	Glaciers at intervals in the Sierra. Lakes in desert basins. Mountain rejuvenation in the Sierra, the south, and the Coast Ranges	Ice advance and retreat four or five times. Loess and morainal deposits	
	Tertiary	Pliocene	Retreat of sea from interior valleys. Land and marine sediments. Volcanism	Cascadian revolution	Rise of Cascades (mountains of central Oregon and Washington)
		Miocene	Extensive advances of sea. Thick marine sediments. Volcanism	Volcanic activity in eastern Oregon and Washington	Wearing down of Sierra and Rockies
		Oligocene	Retreat of sea. Land-laid beds in south	Extensive continental beds in the West	
		Eocene	Extensive flooding of coast and interior by shallow seas	Land sediments in Rocky Mountain area	Local uplifts along California Coast
		Paleocene	Little or no mountain building in California	Continental deposits in Rocky Mountain region. Laramide revolution	
Mesozoic		Cretaceous	Extensive flooding in the north and along southern coast. Islands and peninsulas offshore, continuing into the Pliocene	Great shallow seaway from Gulf of Mexico to Arctic Ocean. World-wide flooding	Rise of Rocky Mountains
		Jurassic	Seaway over present Coast Ranges and great valley	Swamp beds in Rocky Mountain region	Sierra Nevada uplift
			Rise of Sierra Nevada		
			Seas over eastern California and Sierra Nevada	Desert sandstones in Southwest	

CHART 2. Geological events are shown in brief outline; those relating to California are compared with general conditions elsewhere in North America. The record is largely one of mountain-building movements caused by crustal warping, faulting, and collapse. Such recognizable changes are used to separate the main subdivisions of the time scale. Incursions of the sea over the land and regressions of the sea have left their records in the alternating marine and continental sediments—distinguishable by the fossils they contain. Such transgressions also mark subdivisions of the scale. The advance and retreat of the northern ice sheets during the Pleistocene are recorded by morainal material dumped by the ice, and by wind-blown deposits of rock flour, called loess. Such deposits delimit the various stages in which the ice crept down over Canada and the north central states. The names Nebraskan, Kansan, Illinoian, and Wisconsin, in column 2, refer to the glacial

166

CHART 2. (cont.)

	Triassic	Warm clear seas across northeastern California: Shasta region, Sierra region, Inyo Mountains area	Palisade disturbance. Lowland flood plains in Rocky Mountain region. Many volcanoes. Land sediments along Atlantic Coast	
Paleozoic	Permian	Seas much reduced in the West. Clear seas over extreme northern California. Lack of marine sediments indicates that much of California lay above sea level	Appalachian revolution. Ancient mountains formed in Europe and eastern North America	Appalachian Mountains rise
	Pennsylvanian	Less extensive seaways with small land areas in Klamath area	Extensive coal swamps and lowlands	
	Mississippian	Warm seas across eastern and southern deserts; highlands across Nevada, possibly extending westward to Inyo County	Coal swamps with scale trees and seed ferns. Cooler climates	
	Devonian	Seas in the north and east, possibly with land areas in extreme eastern deserts of Inyo County and central Arizona	Warm climates far to north, with first land plants, amphibians, and seed-fern forests	Acadian revolution
	Silurian	Seas over Shasta and Inyo regions	Development of plated fishes with jaws. Extensive seas over eastern and northern North America	Caledonian revolution
	Ordovician	Continuation of seas in various troughs over eastern areas and southern California	First vertebrates. Extensive seas in east and north	Taconic revolution
	Cambrian	Seas extend over eastern desert areas and most of southern California	Widespread seaways over basin and Rocky Mountain area, north into Canada	Green Mountain uplift
Pre-Cambrian	Algonkian	Complicated sedimentary history, little known in California	Many mountains raised up and worn away	Killarney revolution
	Archean	Old crystalline rocks in Grand Canyon region, and southeastern deserts	Ancient deposits in Canada	Laurentian revolution

stages in north central United States. The Californian mountain glacials—McGee, Sherman, Tahoe, and Tioga—are not known to be synchronous with these.

Mountain-building movements in the Far West are not related to those elsewhere. The Laramide revolution which formed the original Rocky Mountains did not extend into California; and the Sierran, Cascadian, and Coast Range uplifts are not represented in eastern North America.

Transgressions of the sea, which form a varied and intricate part of the paleogeography of California, are often caused by submergence of the land. Submergence occurred on a grand scale during the Cretaceous. It was extensive in space and in time, and it is recorded in thick, fossiliferous marine sediments. Land-laid (continental) beds are much thinner and are poorly represented. Consequently, the fossil marine faunas are better known in California than the ancient land life is.

CHART 3. Stages in the Tertiary land life are shown; names of the California fossil horizons are given in approximate order from older to younger in column 5. Recently adopted names for the North American faunal ages are given in column 3 (69); the position of the California deposits in relation to these is shown in column 4.

Glacial stages of the European Pleistocene appear in column 2. These are Potassium-Argon dates, in millions of years, from Everneden, Savage, Curtis, and James. Opposite them in column 3 are the tentative North American glacial and interglacial equivalents.

CHART 3. Continental Deposits and Fossil Faunas and Floras

Period	Epoch		North American faunal ages	California deposits	California fossil faunas and floras
QUATERNARY	Recent 11,000		Postglacial	Asphalt springs, alluvium, gravels, recent burials and caves	Later caves, rockshelters, and mounds
	Pleistocene Fourth Glacial 70,000* Third Glacial 300,000 Second Glacial 700,000 First Glacial	Glacial and interglacial stages	WISCONSIN-MANKATO PEORIAN WISCONSIN-IOWAN SANGAMON ILLINOIAN YARMOUTH KANSAN AFTONIAN NEBRASKAN	Cave deposits in Sierra and Shasta limestones, asphalt springs, alluvial and playa beds, stream terraces, sea terraces, marine and continental beds Basalts and other lava flows	Sierran and Shastan caves McKittrick (Kern) Rancho LaBrea (Los Angeles) Carpinteria (Santa Barbara) Rodeo (Contra Costa) Manix (San Bernardino) Irvington (Alameda)
TERTIARY	M.Y. 1.5 ... 3.5		Blancan	Clays, sands and gravels, stream and pond deposits	Coso Mountains (Inyo) Pittsburg—Tehama San Timoteo (San Bernardino) San Joaquin clay
	6.4		Hemphillian	Volcanic ash	Petaluma (Sonoma) Pinole tuff (Contra Costa) Mulholland, Kern River
	9.9 Pliocene ... 11.7		Clarendonian	Stream, lake, and ash deposits in Coast Ranges and southern Sierra	Upper Mint Canyon (Los Angeles) Ricardo—Upper Chanac (Kern) Siesta (Contra Costa) Ingram Creek (San Joaquin) Black Hawk (Contra Costa) Orinda—Lower Chanac
	12.3 ... 15.6		Barstovian	Barstow beds, stream gravels and sands	Barstow (San Bernardino) Cache Peak (Kern) Coalinga (Kern)
	Miocene		Hemingfordian	Tehachapi sandstone	Phillips Ranch (Kern)
	21.3 25.6		Arikareean		Tecuya (Kern)
			Whitneyan	Upper Sespe	South Mountain (Ventura) Kew quarry (Ventura)
	Oligocene 31.5		Orellan		
	33.4		Chadronian		Titus Canyon (Inyo)
			Duchesnean	Lower Sespe	Sespe Creek Pearson Ranch } (Ventura) Tapo Ranch Poway (San Diego)
	45.0 Eocene 49.0		Uintan		
			Bridgerian		
	49.2		Wasatchian		
	Paleocene 64.8		Tiffanian Torrejonian Puercan		

*G. Brent Dalrymple, "Potassium-argon dates of three Pleistocene interglacial basalt flows from the Sierra Nevada, California." *Bull. Geol. Soc. Amer.*, vol. 75, Aug. 1964, 753–757.

CHART 4. Names and ages of marine beds in California (adapted from C. E. Weaver *et al.* [66], who also provide lists of guide fossils). It is hard to determine the exact age relationships of the marine and continental faunas. This is more difficult far inland than it is along the coast, where fossil-bearing marine and continental beds may overlap. Overlapping establishes the marine-continental correlation in the Pliocene Jacalitos and Etchegoin formations of the Coast Ranges and the valley of the San Joaquin. The wavy lines represent unconformities: intervals not represented by sediments.

CHART 4. Later Marine Formations in California

Era	Period and Epoch		South Coast Ranges	North and Central Coast Ranges	Interior
Cenozoic	Quaternary	Recent	Palos Verdes Los Angeles basin San Pedro Ventura basin Saugus	Terraces Tulare formation Millerton formation	Lake Coahuila
		Pleistocene			
	Tertiary	Pliocene	Pico San Diego formation Repetto	Paso Robles formation Pinole tuff Tassajara	San Joaquin clay Etchegoin—Kern River Jacalitos
		Miocene	Santa Margarita sandstone Sisquoc Modelo "Monterey" Rincon Temblor Vaqueros	San Pablo Briones Monterey shale San Ramon	Shark Tooth Hill (Kern)
		Oligocene	Marine Upper Sespe	Kirker	Pleito. Wheatland
		Eocene	Tejon Domingine. Rose Canyon Lower Sespe Capay	Markley Kreyenhagen	Reeds Canyon Ione formation
		Paleocene	Meganos (restricted) Martinez	"Meganos" Martinez	
Mesozoic	Cretaceous		Trabuco conglomerate, Santa Ana Mountains. Narrow area along coast, from San Diego to Ventura County	Moreno Panoche ~~~~Orogeny~~~~ Pacheco ~~~~Disturbance~~~~ Horsetown Paskenta ~~~~Upwarping~~~~	Chico group Shasta group
	Jurassic		Franciscan series	Knoxville Franciscan	Sierran granites
			Marine sediments, metamorphics, and volcanics		Mariposa slates
	Upper Triassic		Santa Ana limestone	Shasta limestone	Inyo limestone

CHART 5. Animals on the left, plants on the right. The Pre-Cambrian history is almost unknown. The emergent evolution of life depends principally on a long series of primary inventions, called "basic patents" by Professor William K. Gregory (20). These permitted greatly increased complexities, and galaxies of innovations. Some of the outstanding basic patents: invention of cell chemistry, i.e., the controlled reactions of amino acids held in an organized system, the cell nucleus; invention of chlorophyl to enable the plant to utilize sunlight; development of multicellular organisms with special sex tissues undergoing accurate division (mitosis) of hereditary nuclear materials; development of roots, leaves, sap flowing through conducting stems, and finally wind-borne pollen in land plants; development of flowers and insect-borne pollen in flowering plants; use of soluble salts to build shells, plates, spicules, and to support complex organs in animals, with invention of nervous and muscular tissues for sensation and movement, blood in closed body spaces and kidney cells to eliminate waste products from the blood, and balancing apparatus to direct movement; growth of light-sensitive areas (pigment spots) in the skin of invertebrates, leading finally to development of retina, eyeball, and lens; development of pigment spots inside brain of primitive transparent chordate, permitting accurate direction of movement in respect to light and presaging the final inverted form of the vertebrate eye; concentration of brain and sense organs in a head; development of a central, stiffening notochord and an internal skeleton as a base of support for muscles in rapid, well-directed swimming of early chordates; invention of muscular rudder fins and swimming tail; transformation of skin spines to form teeth on vertebrate jaws, permitting predacious habits; rise of air-breathing lungs in stagnant-water fishes, leading to eventual land life; change of fins to limbs and feet; breeding on land, and active locomotion on land and in air, by reptiles; control of body temperature (warm blood) in birds and mammals, together with insulating and protecting hair and feathers; invention of the mammalian placenta, providing longer and more secure embryonic life in placental mammals; increased complexity of brain, intelligent care of young; erect posture, freedom of hands to hold weapons, increased length of juvenile life with longer period of reproduction and learning and increased effectiveness of intelligence in man; ability to manufacture implements and to organize social life, with preservation of culture and discoveries leading to group survival for many generations.

CHART 5. History of Life

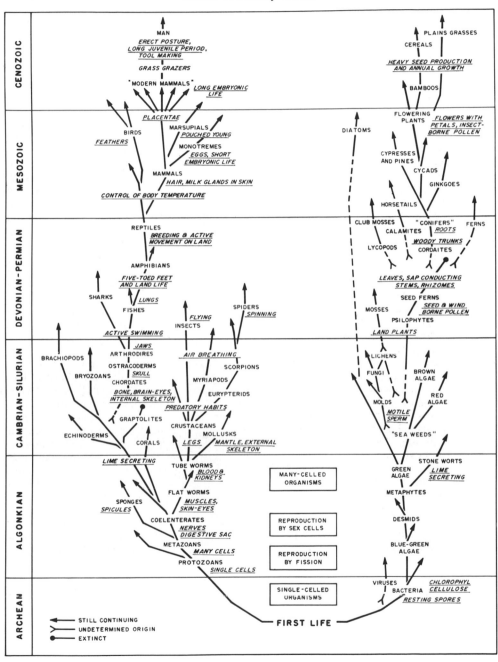

CHART 6. Datings and sequences of the California tar pits are uncertain. The common occurrence of *Bison antiquus* at McKittrick, at Rancho La Brea, and in the Bignell loess of Nebraska (53) seems to establish a post-Tazewell and pre-Mankato age for the older tar pits. Carpinteria is hesitantly placed in the immediate post-Tazewell mainly because of the presence there of forests of the Monterey type, 250 miles south of where they typically occur today. It is possible that remnants of such forests remained in the Santa Ynez Mountains until Tioga time, since they still grow on the near-by islands. If so, Carpinteria could be much later than Tazewell. It does not seem to have been contemporaneous with the hot, dry periods represented in the older Rancho La Brea, for its position, only 100 miles north, could hardly account for its northern trees and birds. High intervening mountains might have created climatic barriers, but so late an orogeny is not evident. A Rancho La Brea climate, like that of northern Lower California, so close to a forest of the Monterey type, seems hardly possible. The Carpinteria pits were probably not active at the same time as those of the early Rancho La Brea.

The correlation of the Tioga with the Mankato, the latest phase of Wisconsin glaciation, agrees with observations by Matthes on the nature of the fresh glacial pavements and morainal weathering since Tioga time — estimated "conservatively" by Matthes at 10,000 years, and confirmed by radioactive carbon (C^{14}) datings.

The date of the end of European glaciation (Finno-Scandian moraines) is placed at 10,000 years by the Scandinavian varve counts. Varve counts in North America by Antevs and others have not been brought down to the present day; the dates are therefore subject to correction. Antevs' studies have placed Lake Mojave in a pluvial period extending from about 30,000 to 10,000 years ago. The California climates as indicated by faunas and floras do not seem to fit too well into this long pluvial period; Lake Mojave might therefore be regarded as a result of ice melt of the Tiogan glaciers, and as a part of the last general flooding of the southern desert basins. Antevs has recently (4b) proposed the terms "anathermal," "altithermal," and "medithermal" for post-Pleistocene climates of the Southwest. Late views concerning the extent of these wet and dry periods are indicated in chart 7.

The Folsom type site contains *Bison taylori*, which seems to be of post-Wisconsin age. Folsom artifacts in Sandia Cave, near Albuquerque, lie above Sandia artifacts and are separated from the Sandia levels by an aqueous deposit of ocher — possibly the time equivalent of the Mankato glaciation farther north.

The Shasta caves, Potter Creek and Samwel, contain remains of peculiar wild cattle *(Euceratherium)*, and a short-faced bear. Related forms have been found at McKittrick and in Burnet Cave, New Mexico. In Potter Creek Cave a mountain goat has been found. The age may be as early or earlier than the Tioga glaciation, but in Samwel Cave the thin travertine deposits covering some of the bones might indicate a moderate or short, wet period, rather than the whole of Tioga time. Hawver Cave, in the Sierra, contains well-cemented bone breccia, and its fauna may be slightly younger than that of the Shasta caves. Bones of the Pleistocene bison *(B. antiquus)* have been found there. Caves such as the Sierran Murphy's and Moaning Cave, which have much unconsolidated refuse, must be postglacial.

References to late Pleistocene faunas may be found in *Bibliographies of Fossil Vertebrates,* Geological Society of America, Special Papers 27 and 42 and Memoir 37, and in Sellards (54).

174

CHART 6. Late Pleistocene Chronology in the Southwest

Years ago, estimated	California	Artifacts	Great Plains and Southwest	Bison	Events
11,000	Tioga glacial		Mankato glacial		Wet period
	Hawver Cave		Sandia and Manzano, New Mexico	Bison antiquus	First Man in America (Sandia)
	Shasta caves		Burnet Cave, New Mexico		*Euceratherium* with mountain goat and short-faced bear
	McKittrick tar seeps Lakes and alkaline flats				Abundant mammalian faunas
		No evidence of human occupation			Desert lake beds dry
	Rancho La Brea older pits 15,000*				"Cool" coastal desert along southern California coast
	Varying climates largely warm and dry		Bignell loess		Recession of forests along coast of southern California
	Carpinteria tar pits?				Lake Lahontan deeply filled
25,000			Tazewell glacial		Monterey-type forests on coast of southern California
50,000			Peoria loess	Long-horned bison	
			Iowan glacial		

*Science, vol. 131, no. 3402, 1960, 712–714.

CHART 7. Human habitations in the Southwest have been dated as far back as two thousand years by A. E. Douglas' method of tree-ring counting. The Mayan calendar gives a little longer sequence of recorded dates in Mexico. Between these recent datings and the first appearance of man (Sandia) was a long sequence of cultures in the Southwest.

Important key sites are the Leonard Rockshelter and Gypsum Cave in Nevada and Ventana Cave in Arizona. Leonard Rockshelter in Pershing County, Nevada, covers a long stratified sequence of dated events (27c). At Gypsum Cave (23), deposits indicating an early wet phase — sands, gravels and travertine — are covered by dry debris with pre-Basketmaker darts and other artifacts associated with ground-sloth, horse, and camel remains. At Ventana Cave, Haury (24) has discovered a sequence of cultures: an early type (Sulphur Spring) with stone spear points in cemented volcanic breccias, over which were laid unconsolidated deposits ranging through two widely separated stages of Cochise culture to Hohokam culture with pottery. Early Ventana and Clovis artifacts may be related to the Folsom points; Clovis contains a bison doubtfully identified as *B. taylori*.

Pumice-covered sites in the Oregon caves also contain evidence of early cultures associated with horse, camel, and elephant. The geological evidence points to an age of about 6,000 years, before the great pumice eruption and collapse of Mount Mazama that formed Crater Lake (14, 15). This has been confirmed by the carbon-14 method of dating (6,400 years).

Borax Lake and Los Angeles man are of doubtful age — the latter was a deeply buried human skeleton in a layer which contained mammoth remains — but an actual association has not been established. The Tranquility site near Tulare Lake contains heavily petrified human skeletons in probable association with camel jaws (29).

References to climatic history may be found in Antevs (4a, 4b), and in Sauer (52a, 52b); and to California occurrences, in Heizer (27a, 27b). Dates given in italics have been determined by the carbon-14 method, a new and promising means of dating organic material of less than 20,000 years of age. The estimates depend upon the persistence of radioactive carbon (C^{14}) in such remains. It is postulated that cosmic-ray neutrons, entering the atmosphere from outer space, react with the nitrogen in the upper atmosphere to produce various forms of carbon and radiocarbon. Some of these substances may be absorbed during the carbon dioxide intake of organisms, both marine and terrestrial; they are considered to remain in constant equilibrium in the atmosphere and in living things, and will enter into the carbonates of sea water. The half life of C^{14}, the most persistent of the radioactive carbon isotopes, is calculated at $5,568 \pm 30$ years. Furthermore, the rate of its measurable emanations is probably not affected by chemical and physical changes in its environment, although the dating of wet and otherwise altered and adulterated material is still questionable.

The C^{14} method of dating was originated at the University of Chicago and has been tested there on historic remains of known age (5a). Datings are now being extended to late glacial and postglacial materials (5b). Present results seem to show that the last great advance of the Wisconsin glaciation in America (Mankato stage) corresponds in age with the final advance of the ice across northern Europe (12,000–10,000 years ago). There are some remarkable concordances with other

CHART 7. Recent Chronology in the Southwest

Years ago	California and Oregon	Nebraska, Colorado, Texas, New Mexico, Arizona, and Nevada	Mexico, Central and South America	Events
100	U.S.A.	Occupation	Mexico and southern republics	Dwindling of mountain glaciers
130	Russia	Mexico		
450	Spain	Tree-ring dates — Pueblo V	Spanish Occupation	
	California Indians. Younger shell mounds and settlements.	650 — Pueblo IV — Hohokam	Aztec — Toltec	Great drought in Southwest
1,000		950 — Pueblo III — IV		
	Emeryville	1,150 — II — III		Shallow filling of desert lakes at intervals
		1,250 — I — IIb — Pottery	Maya	
		1,550 — Basketmaker III — IIa		Rebirth of small mountain glaciers
2,000	Orwood, Ellis Landing	2,000 — Basketmaker II — I — Pottery		
	Oldest shell mounds	San Pedro (Cochise), 2,400 / Lovelock, Nevada, 2,500 / Leonard Rock IV, 2,700	Beginnings of recorded Mayan history	
3,000			Olmec 3500–2100	
	Oak Grove	Moderately wet		
4,000	Borax Lake (?) Windmiller settlements near Lodi, 4,050	Chiricahua Stage (Cochise), 4,000 / Moister	Ancient pottery sites / Tepexpan man, 4,000	Disappearance of desert lakes and mountain glaciers
	Topanga–San Dieguito?	Abilene and Gibson, Texas, with grinding stones and mammoth		Oldest Kentucky mounds, 4,900
5,000	Los Angeles man (?) Lower Klamath Lake (?)	Dry climates	Beginnings of agriculture ?	Last mammoths
		Leonard Rock III, 5,700		
6,000	Oregon Caves, antedating collapse of Mt. Mazama, 6,400			Destruction of Mt. Mazama— formation of Crater Lake
7,000	Tranquility (?) Pinto Basin (?)	Yuma (Sage Creek), 6,800 / Leonard Rock II, 7,000		Last camels and horses
	Lake Mojave (?)	Sulphur Spring (Cochise), 7,750		Last saber cats and dire wolves
8,000		Wet / Leonard Rock I, Nevada, 8,600	Last ground sloths in S. America / Palliaike Cave, 8,600	Last tapirs and ground sloths in North America
9,000	Fort Rock Cave culture, Oregon, 9,000	Clovis, New Mexico (?) / Wet climates		Filling of desert lake basins in California and Nevada to highest post-Pleistocene levels
10,000		Medicine Creek, Nebraska / Plainview, Texas, culture, 10,000 / Folsom, New Mexico (?)		
11,000	Tioga glaciation	Mankato glaciation		End of "Wisconsin" glacial stage

Vertical labels, California/Oregon–Nebraska boundary (top to bottom): Bow and arrow and mortars; Younger levels, Gypsum Cave; darts; atlatl and; stones; Seed-grinding; Stone spear points and knives; 10,500, Gypsum Cave, older levels

Vertical labels, Mexico column: Bow and arrow; Mayan calendar; Bison bison, extinct to modern subspecies; Bison taylori

Vertical labels, Events column (right margin): Medithermal; Altithermal; Anathermal datings, and some as yet unaccountable discrepancies: historically dated test samples have shown an error of less than 10 per cent.

The original carbon-14 date of the Folsom type site was erroneous because the charcoal used in estimating it was from an unassociated "fire pit." The age of this pit has recently been determined as about 4,500 years, and none of the artifact sites so far analyzed appears to be of preglacial (Pleistocene) age. Ancient, woven, sage-fibre sandals from Fort Rock Cave in Oregon, buried below one of the early pumice eruptions of Newberry Volcano, are dated at about 9,000 years; and the earliest appearance of man in southern South America (Palliaike Cave, southern Chile), associated with an extinct ground sloth and horse, is given as $8,639 \pm 450$ years. The Nevada caves (Gypsum and Leonard) are given a generous age (10,500 and 8,600 years, respectively); the Arizona Cochise sites are given an age of 7,750–2,400 years (5b). The Plainview, Texas, bison quarry has received a carbon-14 date of about 10,000 years. This interesting culture is regarded by its describers as intermediate between Folsom and Yuma (55).

178

Glossary

Algae. — A composite group of simple plants without stems, leaves, or roots. Includes microscopic single-celled forms (blue greens) as well as the pond scums, frog silks, seaweeds, and stoneworts. Algonkian to Recent.

Algonkian. — The period immediately after the Archean and before the Cambrian. Duration about five hundred million years. (Chart 1.)

Amber. — Fossilized resin, frequently enclosing insects, spiders, bits of vegetation, and other small objects. The Oligocene amber from the Baltic Sea is famous as a source of fossil insects.

Ammonites (Ammon's horns). — Coiled and uncoiled, chambered shells, usually snail-like, but with segments separated by sutures. Some reached a diameter of six feet. Some were probably floaters, buoyed by the empty chambers of the shell. The animal itself was similar to an octopus. Its nearest living relative is the chambered nautilus (fig. 13). The early ammonoids (fig. 10) of the Paleozoic were related to the earlier nautiloids (fig. 11) and the later, squidlike belemnites. True ammonites with complicated sutures lived in the Triassic and in the Cretaceous — ammonoids continued into the Triassic. *See also* Ceratite ammonoids.

Aphaneramma. — A labyrinthodont amphibian (stegoceph) with an extremely long, slender skull, from the early Triassic of Spitzbergen; a close relative is found in the Moenkopi Formation of Arizona.

Aplodontia. — The mountain "beaver," or sewellel, of the northwest Pacific coast and the Sierra. A primitive burrowing rodent belonging to one of the oldest of the rodent groups. Teeth simple, peglike; jaw muscles weak. Miocene to Recent.

Archean. — The oldest and longest period of earth history. Duration about two billion years. (Chart 1.)

Arthrodires. — Archaic plated fishes with gills concealed in chambers beneath the head shields and with a joint between the head plates and the neck plates.

Artiodactyls. — An order of mammals, mostly herbivorous, including the pigs, hippos, pig deer (oreodonts), camels, giraffes, deer, antelopes, and cattle, in which the axis of the foot passes between the third and fourth toes; the feet are functionally reduced to these two toes in advanced forms. The ridges of the teeth usually have a pattern. *See also* Perissodactyls.

Belemnites. — *See* Ammonites.

Bison taylori. — A moderately long-horned, extinct species closely related to *Bison antiquus* and to the modern North American *Bison occidentalis*. Found associated with the spear points of Folsom man, who evidently hunted it. Immediate post-Pleistocene.

Blastoids. — Budlike ancestors of the stone lilies, less than an inch in diameter, something like a sea urchin on a stalk. Ordovician to Pennsylvanian.

Brachiopods. — Shelled marine animals resembling clams but not related to them. The two parts of the shell are unlike and are composed of lime or chitin. The feeding "gills" in the spiriferous forms are carried on a spiral skeletal support (spire) resembling a coiled spring. Cambrian to Recent — abundant in the Paleozoic.

Brontosaurus. — Largest of the North American dinosaurs, swamp-living, quadrupedal, long-necked, with a light skull and neck swung from a ponderous heavy body upraised on pillar-like limbs. Upper Jurassic of Wyoming. Related brachiosaurs from East Africa were the largest of land animals — nearly ninety feet in length, standing forty feet high when the head was raised.

Bryozoans. — Moss animals, living in small colonies of tubelike and lacelike form, that cover rocks and shells with thin, delicate perforated crusts. Some of them look like tiny seaweeds. The individual animals are minute polyps distantly related to the brachiopods. Ordovician to Recent.

Calamites. — Ancient plants of the horsetail (scouring rush) group, often growing as large trees with pithy pipestems, jointed like the horsetails but with longer leaves and larger cones. Carboniferous and Permian, possibly into the early Triassic. *See also* Horsetails.

Cambrian. — The first period of the Paleozoic era. (Chart 1.)

Carboniferous. — The Age of Amphibians and Coal, comprising the subperiods Mississippian and Pennsylvanian. Duration about 105 million years. (Chart 1.)

Cenozoic era. — The Age of Mammals, Paleocene to Recent, inclusive. Duration about seventy million years. (Chart 1.)

Ceratite ammonoids. — Ammonoids with moderately complex sutures (fig. 10).

Chert. — A glassy, silicious rock often formed in hot alkaline waters, sometimes composed of the skeletons of radiolarians or of glass sponges.

Chirotherium. — A mud-walking reptile, known only from numerous tracks in the Triassic of Europe, North America, and South America. The hind feet were broad, with heavy, hooflike claws and thick toes in the later forms. The fifth toe turned outward like a thumb, giving the hind footprint the appearance of a hand. The front feet, much smaller, were not much used for walking in some of the latest forms. Relatives of these animals probably gave rise to the early bipedal dinosaurs.

Chumashius. — An Eocene primate of southern California (pl. 6), related to the modern tarsiers of the Orient and, more distantly, to the lemurs of the Eocene.

Club mosses. — Small surviving forms, sometimes called spike mosses and ground pines, are mosslike plants that often grow in dry regions; they bear spores in clublike cones on terminal twigs, and have tiny leaves spirally arranged like scales on the stem. They grew as large trees (lycopods) during the latter Paleozoic and were especially abundant in the coal forests of the Carboniferous. Devonian to Recent.

Cotylosaurs. — The most primitive of the reptilian subclasses, directly descended from the amphibians, characterized by solid-roofed skull, short neck, awkward limbs and feet. Pennsylvanian to Recent (turtles). Most of them died out in the Permian and early Mesozoic.

Creodonts. — Ancestral carnivorous mammals of the Lower Tertiary (pl. 6), smaller-brained and with less effective teeth and feet than modern dogs, cats, weasels, and other descendant forms. Paleocene to Pliocene.

Cretaceous. — The last period of the Mesozoic. Duration about fifty million years. (Chart 1.)

Crinoids. — The stone lilies of the sea, animals with flower-like heads and often with stalks made up of circular discs. Related to the starfishes and sea urchins (Echinodermata). Early forms probably motile, later ones sessile.

Cro-Magnons. — Well-advanced cave men of the last glacial, reindeer stage in Europe. Similar in stature and brain development to men of today. Late Pleistocene, 15,000-30,000 years ago.

Crossopts. — Crossopterygians, fishes with fringe fins, fins on stumpy "legs," nostrils perforating the skull evidently for breathing into a lung. Ancestral to the Amphibia. Devonian to Recent; only one living species has been described.

Cycads. — Related to the pines and ginkgos, with solid trunks externally roughened like a pineapple, and palmlike leaves unrolling like fern fronds; intermediate between ferns and flowering plants (angiosperms). The cones bear flowers and seeds. The modern ornamental types are known as "sago palms." Triassic to Recent, abundant in the Jurassic.

Cystoids. — Nutlike, short-stalked ancestors of the blastoids and stone lilies. Cambrian to Devonian.

Desmostylus. — An extinct genus of peculiar sea mammals which lived in the Miocene along the coasts of the North Pacific from Lower California to Japan. These had long upper and lower jaws, with tusks; and some of the Japanese and California forms apparently had stout limbs, large flat feet, and big tusks, suggesting distant relationships to the mastodonts.

Devonian. — The fourth period of the Paleozoic era. (Chart 1.)

Diatoms. — Microscopic, single-celled plants (algae), with transparent silicious shells fitting around the body like a pillbox and ornamented with geometrically accurate tracery (fig. 17). Regarded as an important source of petroleum. Deposits of masses of these shells have been called diatomite. Diatoms occur in fresh water and damp soil, as well as in the sea. Cretaceous to Recent.

Diatryma. — An extinct, flightless, predatory bird as large as an ostrich and with a huge head and beak. Eocene of Wyoming.

Dicynodonts. — A group of herbivorous, mammal-like reptiles (therapsids), with turtle-like beaks, probably horny, and usually toothless except for a pair of tusks (canine teeth), which are sometimes absent (pl. 3). Common in the Upper Permian of South Africa and Eurasia, ranging into the Triassic of North and South America.

Dinichthys. — A genus of large, predatory, arthrodiran fishes, armor-plated, heads jointed across the top of the neck, jaws with heavy, piercing teeth. Upper Devonian.

Dinosaurs. — A general term which refers to many diverse extinct reptiles in two main groups. The early types were small, two-legged, active, with small heads, long necks, and long hind legs. Ponderous quadrupedal giants lived in the Upper Jurassic, and in the Cretaceous along with many curious armored forms. The fossil duck-billed dinosaur recently discovered in California belongs to a group common in the Rocky Mountain Cretaceous, characterized by the masses of teeth, which are fused together into millstones for chewing vegetation. Middle Triassic to end of Cretaceous.

Diplodocus. — A dinosaur similar to *Brontosaurus* and more slender.

Edentates. — A diverse assemblage of primitive mammals belonging to at least three orders, not closely related. They usually had simple teeth without enamel, or were toothless. They were slow-moving, digging and climbing, insectivorous and herbivorous. Lower Eocene to Recent.

Eohippus. — The earliest genus of the horse family (Equidae), ranging in size from a cat to a small sheep; forest-living browsers, ancestral to all later perissodactyls. Basal Eocene.

Ephedras. — Shrubby plants belonging to the order Gnetales, bearing small cones and distantly related to the pines and cycads. Only three living genera of the Gnetales exist: the Mormon tea, or squaw tea *(Ephedra),* of the Southwestern deserts and Asia, from which the drug ephedrine is obtained; *Gnetum,* a widely distributed tropical plant; and *Welwitschia,* a desert plant of Angola and South-West Africa having two continuously growing strap leaves which arise from a low, cup-shaped, woody "trunk." The only fossil representative of the Gnetales is the genus *Schilderia* from the Upper Triassic of Arizona. This was a tree with a woody trunk reaching about a foot in diameter.

Episcoposaurs. — Heavily armored reptiles with cowlike horns on their shoulders and spiny plates along the sides of the body and tail, distantly related to phytosaurs, crocodiles, and dinosaurs (pl. 3). Upper Triassic of Texas and Arizona.

Eurypterids. — Ancient, predatory, water-living relatives of the scorpions. Some reached a length of nine feet. From Cambrian to Silurian they lived in the ocean, after that in brackish water and finally in fresh water in the Pennsylvanian.

Foraminifera. — Single-celled (protozoan) animals usually bearing a limy shell perforated with tiny holes (foramina) (figs. 18, 27). Cambrian to Recent — abundant from Pennsylvanian to Recent, and useful in dating and mapping subsurface formations in oil-bearing rocks. Called "forams" or, more properly, foraminifers.

Fossil. — Any trace of life preserved in rocks older than the Recent epoch. Commonly, the petrified hard parts of organisms, such as bones, teeth, shells, wood, and leaves. A fossil is not necessarily petrified and may not always represent an extinct species.

Ganoids. — General term for fishes having porcelain-like ganoine, a hard surface, on their scales. Sturgeon, garpike, and bowfin are living examples. The ganoids began in the Devonian, were abundant in the Mesozoic, and now are much reduced in numbers.

Glyptodonts. — Large tortoise-like mammals, South American edentates, with domed, ornamental armor over head and body and round the tail; like armadillos but more ponderous. Eocene to Pleistocene of South America, reaching North America and California in the late Pliocene.

Gomphotherium. — Long-jawed mastodonts of the California Pliocene, with elephantine bodies but with short legs and with tusks in the lower as well as the upper jaw (end papers). The downturned lower jaws may have been used to dredge vegetation in streams and lakes.

Graptolites. — Extinct colonial polyps united by branching rods and filaments of chitin, horny inert material (fig. 7). Many were equipped with floats to buoy them on the surface of the sea. The name ("rock writing") was given because the fossil filaments resemble the cuneiform inscriptions on Chaldean clay tablets. Usually classed with the coelenterate hydrozoans, but relationships to the bryozoans and, recently, to the ancestors of the vertebrates (hemichordates) have been suggested. Ordovician and Silurian.

Ground sloths. — Ponderous ground-dwelling relatives of the South American tree sloths (edentates). Their skin was thick and often contained very hard, embedded bones about the size of small marbles; their hair was shaggy and coarse and, like that of the tree sloths, was infested with green algae. All are now extinct. Pliocene to late Pleistocene and Recent.

Hadrokkosaurus. — A stegoceph with extremely broad, short skull and huge orbits, from the Moenkopi Formation of Arizona, middle Triassic, related to a similar genus from South Africa (67c).

Hipparion. — See *Neohipparion.*

Horsetails. — Scouring rushes, hollow-stemmed, with joints that pull apart like a fishing pole. The leaves are inconspicuous. The green stems are roughly ridged and are abrasive because of the silica they contain. Distantly related to the club mosses and ferns. Horsetails today are small plants, related to the much larger extinct calamites. Triassic to Recent. *See also* Calamites.

Hydrozoans. — Delicate marine animals commonly growing in clusters or colonies, consisting of polyps encased in gelatinous cups and often secreting limy platforms (coral) for support. Related to the jellyfishes (Coelenterata). Algonkian to Recent.

Hyraces. — Animals of the hyrax groups (order Hyracoidea), distantly related to the elephants. The modern forms, living in Africa and Eurasia, resemble rabbits; they are the coneys of the Bible. Ancient ones were as big as bears. Oligocene to Recent.

Insectivores. — Mammals of the order Insectivora, including moles, shrews, desmans, elephant shrews, European hedgehogs, and diverse relict forms not closely related. The most ancient group of placentals. Cretaceous to Recent.

Jurassic. — The second period of the Mesozoic. Duration about thirty million years. (Chart 1.)

Lake Mojave. — An old lake bed now trenched by the Mojave River in San Bernardino County, California. By ruling of the United States Geographic Board, the names of the river, town, valley, sink, and desert (in the Great Basin) are spelled *Mojave;* the spelling *Mohave* is used for the names of the mountains, canyon, rock, wash, Indians, and Indian Reservation on the Colorado River drainage. Archaeologists named the site Lake Mohave, but the spelling Mojave is used here, in order to conform with the ruling of the Geographic Board.

Laramide revolution. — The orogeny, or mountain-building movement, at the end of the Cretaceous in the Rocky Mountain region.

Lion jaguar. — The "lion" of the California Pleistocene, found at Rancho La Brea, has lately been regarded as a large-headed jaguar.

Lungfishes. — Dipnoan fishes with a lung as well as gills, with large fan-shaped teeth, one tooth on each side of upper and lower jaws; able to survive in stagnant water. Three living genera, widely dispersed, are relicts of much more abundant dipnoans of the Paleozoic and Mesozoic.

Lycopods. — *See* Club mosses.

Marsupials. — A subclass of mammals, including the kangaroos and opossums, having a brief embryonic period and carrying the premature, larval young in a brood pouch where they are permanently attached to the teats until well developed. Cretaceous to Recent.

Mastodonts. — Simple-toothed elephant-like beasts, with no infolding of enamel bands in the teeth, frequently with tusks in lower as well as upper jaws; short-legged, long-headed, ponderous browsers. Miocene to Pleistocene; exterminated within the last six thousand years.

Mesozoic era. — The Age of Reptiles: Triassic to Cretaceous, inclusive. Duration about one hundred and sixty million years. (Chart 1.)

Miohippus. — An early member of the horse family that still retained primitive short teeth and three-toed feet. Oligocene.

Mississippian. — The lower part of the Carboniferous period. (Chart 1.)

Monotremes. — An aberrant subclass of mammals, including only the living spiny anteaters (*Echidna*) and the duck-billed platypus (*Ornithorhynchus*) of the Australian region. They lay eggs and have many reptilian characters. Fossil history not yet known except in the Pleistocene.

Mosasaurs. — Sea lizards, paddle-footed, long-bodied, with long swimming tails, and flexible jaws armed with pointed or blunted teeth (pl. 5). Fossils of adults are common in the marine chalk beds of Kansas and Belgium, but young skeletons have never been found. Upper Cretaceous, world-wide in distribution.

Nautiloids. — *See* Ammonites.
Neohipparion. — Light-limbed, slender-toothed antelope-like horses retaining remnants of the side toes as well as the main middle hoof. Pliocene.

Ordovician. — The second period of the Paleozoic era. (Chart 1.)
Oreodonts. — A varied and abundant group of extinct piglike deer, clumsy-footed in comparison with antelopes and deer, and with heavier heads and shorter necks. Sometimes called pig deer. Teeth of ruminant type, more deerlike than piglike. Late Eocene to Pliocene.
Ostracoderms. — Earliest of the fossil vertebrates, plated, scaled, and spined, sluggish, sea-bottom feeders, without jaws and with pouched gills; brain and other organs as in the modern lamprey eels and hagfishes. Ordovician to Devonian.

Paleozoic era. — Cambrian to Permian, inclusive. (Chart 1.)
Pennsylvanian. — The upper part of the Carboniferous period. (Chart 1.)
Perissodactyls. — An order of herbivorous mammals including horses, tapirs, rhinos, chalicotheres, and titanotheres, in which the axis of the foot passes through the third toe and the main weight is carried on this toe in advanced horses. The ridges of the teeth are arranged in the form of the Greek pi (π). Eocene to Recent. *See also* Artiodactyls.
Permian. — The last period of the Paleozoic. Duration about sixty million years. (Chart 1.)
Petrifaction. — A petrified object, either fossil or nonfossil. Organic remains which have been infiltrated, and (or) replaced, molecule by molecule, with hard matter — "turned to stone."
Petrification. — The process of alteration or replacement of organic remains by salts or minerals.
Phytosaurs. — Crocodile-like, predacious reptiles (pl. 3) with monstrous jaws, and with nostrils on top of the head instead of at the tip as in crocodilians. Triassic of Eurasia and North America.
Placentals. — A subdivision of the mammals, comprising the great bulk of the modern forms. The fetus is equipped with special membranes (afterbirth) which not only attach it to the uterine walls but enable it to feed, excrete, and respire by transfer of fluids from the mother's blood. The embryo is able to reach a more advanced stage of development before birth than in the marsupials and monotremes. Cretaceous to Recent.
Placerias. — A large North American dicynodont reptile (pl. 3), an immigrant belonging to a group common in Africa and Eurasia from Upper Permian to Middle Triassic. Upper Triassic of Arizona.
Pleistocene epoch. — All of the Quaternary except the brief Recent epoch. (Chart 3.)
Plesiosaurs. — Sea reptiles with broad bodies, long flippers, short tails. Some have very long necks and small heads (California elasmosaurs) and some have large heads and short necks (pliosaurs). Triassic to end of Cretaceous, world-wide; California forms occur in the Upper Cretaceous (pl. 5, 67a).
Pre-Cambrian. — Earth history before the Cambrian, often divided into the Archean and Algonkian, about two and a half billion years. (Chart 1.)
Proboscideans. — Elephants and their relatives including the mammoths and mastodonts. An order (Proboscidea) of mammals with a pair of incisor teeth forming tusks (often present in both jaws) and a battery of enlarged grinding teeth. The jaws are short in later forms: in the elephants the teeth have become exceedingly heavy and have several yards of infolded enamel, providing for long wear and longevity. Upper Eocene to Recent.
Pteranodon. — *See* Pterodactyls.
Pterodactyls. — Batlike flying reptiles, with membranous wings supported by a much-enlarged fifth finger, hind limbs weak with claws like those of bats, evidently for hanging head downward (pl. 5). They were distantly related to dinosaurs and birds. Some had long rudder tails, others were stub-tailed. Some had many fine teeth, probably for catching insects, others like the giant *Pteranodon* were toothless and probably caught fish. The long bones were light and hollow. Jurassic to end of Cretaceous.

Quaternary period. — Approximately the last million years of earth history, including the Pleistocene (ice ages) and the Recent (the last 10,000 years or so). (Chart 3.)

183

Radiolaria. – Microscopic, single-celled (protozoan) animals having a transparent glassy skeleton of great delicacy (fig. 4). Mostly found in depths of the sea. Pre-Cambrian to Recent.

Recent. – The epoch of the last ten thousand years since the retreat of the last glaciation in the Northern Hemisphere. (Charts 3 and 7.)

Rodents. – An order of mammals, including rats, squirrels, and beavers. Rodents feed principally on grain, nuts, fruit, leaves, grass, and bark. They have a pair of gnawing incisor teeth above and below, well separated from the grinding teeth. Paleocene to Recent.

Saber cats. – The largest of these *(Smilodon)* is common at Rancho La Brea; smaller relatives occur in the Tertiary of California. Short-tailed and catlike, with long canine teeth, often protected by flanges on the lower jaws. Usually called saber-toothed tiger.

Seed ferns. – Extinct fernlike plants; some grew as trees, carrying spores or seeds in nutlike pods on their fronds. Regarded as ancestral to the pines and cycads and classed with these (as gymnosperms) by some botanists.

Silurian. – The third period of the Paleozoic era. (Chart 1.)

Smilodon. – The large saber-toothed tiger of the Pleistocene of North and South America (pl. 9). A descendant of the Tertiary saber cats and not closely related to either the lion or the tiger.

Stegocephs. – Stegocephalian amphibians, differing from modern salamanders in having scales, and a solidly roofed skull (pl. 3). There were many kinds, differing more in the skull shape and the construction of the vertebrae (backbone) than in body form. Upper Devonian to end of Triassic.

Stegosaurs. – Peculiar dinosaurs, standing on four legs, small-headed, with rows of large upright plates forming V-shaped crests on back and tail. Upper Jurassic of Europe and North America.

Stoneworts. – Simple plants (algae) which secrete lime in laminated masses to form reefs, tufa domes, and other limestone bodies.

Stromatoporoids. – Colonial hydrozoans related to jellyfishes and corals (coelenterates), which form large globular, concentrically thin-layered, onion-like rock masses. Ordovician to Devonian.

Tamiosoma. – Gregarious marine barnacles living in stony clusters of cylindrical shells, layer upon layer, occupying only the upper levels of the tubular columns at each stage of construction. Miocene and Pliocene of California.

Tertrema. – A long-beaked, stegocephalian amphibian from the Lower Triassic of Spitzbergen. Probably a fisheater, which possibly lived along shore in shallow seas. A similar form occurs in the Lower Triassic of Arizona in land-laid beds.

Therapsid reptiles. – A large and varied group of reptiles descended from the Lower Permian pelycosaurs and approaching the mammals in their structure. Late Triassic therapsids gave rise to the mammals. Upper Permian to Jurassic, chiefly South Africa.

Titanotheres. – Heavy-headed and bulky mammals distantly related to the horses and rhinos, but reaching a much larger size (pl. 7). The later forms had bony bosses and bifurcated horns on the nose, for fighting. The teeth were broad and blunt, an indication of browsing habits. The feet were stubby and elephantine in the larger forms. Eocene and Oligocene.

Tremarctotherium. – The giant, short-nosed, cave bear of the California Pleistocene.

Triassic. – The first period of the Mesozoic. Duration about fifty million years. (Chart 1.)

Trilobites. – Extinct relatives of the crabs, shrimps, and sowbugs (Crustacea) (fig. 6). Body in three longitudinal zones (trilobed) and with many segments and legs. Well-developed eyes and sensory antennae. Covered with hard chitin, which was shed as in the crabs. Sea-bottom feeders and scavengers. Abundant, Cambrian to Silurian; then rapidly decreased.

Trilophosaurus. – A late survivor of a group of slender-limbed, bony-cheeked reptiles (araeoscelids) of the Lower Permian of North America. Triassic of Texas.

Tyrannosaurus. – A giant, two-legged, carnivorous dinosaur with a large head equipped with a battery of teeth like bayonets. The largest of the land carnivores. Upper Cretaceous of North America.

Uma. – A lizard of the dune sands of the hottest California and Mexican deserts, able to scurry, dive, and swim in loose sand.

References

1. Ahrens, Louis. "Measuring Geologic Time by the Strontium Method," *Bulletin of the Geological Society of America,* LX (1949), 217–266, 10 figs., 2 pls.

1a. Alexander, Tom "The secret of the spreading ocean floors." Fortune Mag., February, 1969.

2. Amsden, C. A. "The Pinto Basin Artifacts," *Southwest Museum Papers,* No. 9 (1935), 33–51.

3. Andrews. H. N., Jr. *Ancient Plants and the World They Lived In.* Ithaca, N.Y., Comstock Publishing Company, 1947. ix + 279 pp., 166 figs.

4a. Antevs, Ernst. "Climate and Early Man in North America," in McCurdy *et al., Early Man.* New York, J. B. Lippincott Company, 1937. 362 pp., 54 figs., 27 pls.

4b. Antevs, Ernst. "Climatic Changes and Pre-White Man," *University of Utah Bulletin,* XXXVIII (1948), 168–191.

4c. Ardrey, R. "African Genesis." New York: Atheneum, 380 pp., illustr.

5a. Arnold, J. R., and W. F. Libby. "Age Determinations by Radiocarbon Content: Checks with Samples of Known Age," *Science,* CX (1949), 678–680, 1 fig.

5b. Arnold, J. R., and W. F. Libby. "Radiocarbon Dates," *Science,* CXIII (1951), 111–120.

6. Bailey, T. L. "Late Pleistocene Coast Range Orogenesis in Southern California," *Bulletin of the Geological Society of America,* LIV (1943), 1549–1568, 2 figs., 2 pls.

7. Blackwelder, Eliot. "Pleistocene Glaciation in the Sierra Nevada and Basin Ranges," *Bulletin of the Geological Society of America,* XLII (1931), 865–922, 31 figs., pls. 22–23.

8. Boutwell, J. M. "The Calaveras Skull," *United States Geological Survey Professional Papers,* No. 73 (1911), 54–55.

9. Brodrick, A. H. *Early Man: A Survey of Human Origins.* London, New York, etc., Hutchinson's, 1948. 288 pp., 21 figs., maps. frontis.

10. Buchsbaum, Ralph. *Animals Without Backbones.* Chicago, University of Chicago Press, 1941. ix + 371 pp., many figs. and photos. (Chapter 27 deals with fossils.)

11. Byerly, Perry. "Comment on: 'The Sierra Nevada in the Light of Isostasy,' by Andrew C. Lawson," *Bulletin of the Geological Society of America,* XLVIII (1938), 2025–2031.

12a. Camp, C. L. "A Study of the Phytosaurs, with Description of New Material from Western North America," *Memoirs of the University of California,* X (1930). x + 174 pp., 49 figs., 6 pls., map.

12b. Camp, C. L. "California Mosasaurs," *Memoirs of the University of California,* XIII (1942), pp. 1–68, 26 figs., 7 pls.

12c. Camp, C. L., and S. P. Welles, "Triassic Dicynodont Reptiles." Memoirs of the University of California (1956) vol. 13, 255–348, illustr.

13. Campbell, E. W. C., *et al.* "The Archeology of Pleistocene Lake Mohave," *Southwest Museum Papers,* No. 11 (1937), 9–118, 3 figs., 57 pls.

13a. Colbert, E. H. *The Age of Reptiles.* 1965. New York: W. W. Norton and Co. 228 pp., illustr.

13b. Colbert, E. H. *Dinosaurs.* 1961. New York: E. P. Dutton, 300 pp., illustr.

14. Cressman, L. S. "Results of Recent Archaeological Research in the Northern Great Basin Region of South Central Oregon," *Proceedings of the American Philosophical Society,* LXXXVI (1943), 236–246, 13 maps.

15. Cressman, L. S., and Howel Williams. "Early Man in South-central Oregon: Evidence from Stratified Sites," *University of Oregon Monographs, Studies in Anthropology,* No. 3 (1940), 53–78, 10 figs., 16 pls., map.

16a. Daly, R. A. *The Floor of the Ocean.* Chapel Hill, University of North Carolina Press, 1942. x + 177 pp., 82 figs.

16b. Daly, R. A. "Meteorites and an Earth-Model," *Bulletin of the Geological Society of America,* LIV (1943), 401–456.

17. Deiss, Charles. "Cambrian Geography and Sedimentation in the Central Cordilleran Region," *Bulletin of the Geological Society of America,* LII, 1085–1116, 10 figs.

18. Emery, K. O., and F. P. Shepard. "Lithology of the Sea Floor off Southern California," *Bulletin of the Geological Society of America,* LVI (1945), 431–478, 3 pls., 1 fig.

18a. Evernden, J. F., D. E. Savage, G. H. Curtis, and G. T. James. "Potassium-Argon dates and the Cenozoic mammalian chronology of North America." 1964. American Journal of Science, vol. 262, 145–198.

19. Gregory, J. T. "Osteology and Relationships of *Trilophosaurus*," *University of Texas Publication* 4401 (1945), 273–359, 11 figs., pls. 18–32.

20. Gregory, W. K. *Evolution Emerging: A Survey of Changing Patterns from Primeval Life to Man.* New York, The Macmillan Company, 1951. 2 vols., xxvi + 736 pp., frontis.; [ii +] 1013 pp., 24 chapters of figs. and plates [the entire second volume].

21. Grinnell, Joseph. *Joseph Grinnell's Philosophy of Nature,* edited by A. H. Miller. Berkeley and Los Angeles, University of California Press, 1943. xv + 237 pp., port., figs., map.

22. Gutenberg, Beno. "Seismological Evidence for Roots of Mountains," *Bulletin of the Geological Society of America,* LIV (1943), 473–498, 3 figs., 3 pls.

23. Harrington, M. R. "Gypsum Cave, Nevada," *Southwest Museum Papers,* No. 8 (1933). ix + 197 pp., 77 figs., 19 pls.

24. Haury, E. W., *et al. The Stratigraphy and Archaeology of Ventana Cave, Arizona.* Tucson, The University of Arizona Press; Albuquerque, The University of New Mexico Press, 1950. xxvii + 599 pp., 118 figs., 65 pls.

25. Hazzard, J. C. "Paleozoic Section in the Nopah and Resting Springs Mountains, Inyo County, California," *California Journal of Mines and Geology,* Report 33 of the State Mineralogist (1938), pp. 273–339, 15 figs.

26. Hazzard, J. C., and C. H. Crickmay. "Notes on the Cambrian Rocks of the Eastern Mohave Desert, California," *University of California Publications, Bulletin of the Department of Geological Sciences,* XXIII (1933), 57–80, 1 fig., 1 pl., 1 map.

27a. Heizer, R. F. "A Bibliography of Ancient Man in California." *Reports of California Archaeological Survey,* No. 2 (1948), 22 pp.; No. 4 (1949), 27 pp., 2 pls. [Now *University of California Archaeological Survey.*]

27b. Heizer, R. F. "The Archaeology of Central California, I: The Early Horizon," *Anthropological Records,* XII, No. 1 (1949). iv + 56 pp., 19 figs., 9 pls.

27c. Heizer, R. F. "Preliminary Report on the Leonard Rockshelter Site, Pershing County, Nevada," *American Antiquity,* XVII, No. 2 (1951), 89–98, figs. 39–43.

28. Heizer, R. F., and Franklin Fenenga. "Archaeological Horizons in Central California," *American Anthropologist,* XLI (1939), 378–399, 2 figs.

29. Hewes, G. W. "Early Man in California and the Tranquility Site," *American Antiquity,* XI (1946), 209–215.

30. Hill, M. L. "Structure of the San Gabriel Mountains, North of Los Angeles, California," *University of California Publications, Bulletin of the Department of Geological Sciences,* XIX (1930), 137–170, 6 figs., pls. 15–20.

31. Hopper, R. H. "Geologic Section from the Sierra Nevada to Death Valley, California," *Bulletin of the Geological Society of America,* LVIII (1947), 393–432, 3 figs., 6 pls.

31a. Jahns, R. H., ed. "Geology of Southern California." Calif. Division of Mines, Bulletin 170, 1954. Issued in parts.

32a. Jenkins, O. P. [*et al.*. "Middle California and Western Nevada," *Sixteenth International Congress, Guidebook 16: Excursion C-1* (Washington, U.S. Geological Survey, Government Printing Office, 1933). v + 116 pp., 19 figs., 18 pls.

32b. Jenkins, O. P. [*et al.*] "Manganese in California," *California Department of Natural Resources, Division of Mines, Bulletin* No. 125 (Sacramento, 1943). 387 pp., 46 figs., map.

32c. Jenkins, O. P. [*et al.*] "Geologic Guidebook along Highway 49 — Sierran Gold Belt: The Mother Lode Country," *California Department of Natural Resources, Division of Mines, Bulletin* No. 141 (Sacramento, 1948). 164 pp., 236 figs., 3 pls., 10 maps.

32d. Jenkins, O. P. ed. "Geologic Guidebook of the San Francisco Bay Counties." Calif. Division of Mines. Bulletin 154. 1951. 391 pp., illustr.

33. Knowlton, F. H. *Plants of the Past.* Princeton, Princeton University Press, 1927. xix + 275 pp., 90 figs., 1 pl.

34. Knopf, Adolph. "Time in Earth History," *Genetics, Paleontology and Evolution.* Princeton University Press, 1949, pp. 1–9.

35. Kroeber, A. L. "Handbook of the Indians of California," *Smithsonian Institution, Bureau of American Ethnology, Bulletin 78* (1925). xviii + 995 pp., 78 figs., 83 pls.

36. Latimer, W. M. "Astrochemical Problems in the Formation of the Earth," *Science,* CXII (1951), 101–104, 1 fig.
37. Lawson, A. C. "The Post-Pliocene Diastrophism of the Coast of Southern California," *University of California Bulletin of the Department of Geology,* I (1893), 115–160, 1 fig., pls. 8–9. *See also* Byerly (11).
38. Macdonald, J. R. "The Pliocene Carnivores of the Black Hawk Ranch Fauna," *University of California Publications, Bulletin of the Department of Geological Sciences,* XXVIII (1948), 53–80, 15 figs.
39. Macgowan, Kenneth. *Early Man in the New World.* New York, The Macmillan Company, 1950. 260 pp., illus.
40. Mason, J. F. "Geology of Tecopa Area, Southeastern California," *Bulletin of the Geological Society of America,* LIX (1948), 333–352, 2 pls.
41. Matthes, F. E. "Geologic History of the Yosemite Valley," *United States Geological Survey, Professional Papers,* No. 160 (1930). vi + 137 pp., 38 figs., 52 pls.
42. Merriam, C. W. "Devonian Stratigraphy and Paleontology of the Roberts Mountains Region, Nevada," *Geological Society of America, Special Papers,* No. 25 (1940). vii + 114 pp., 7 figs., 16 pls.
43. Merriam, J. C. *The Living Past.* New York, Charles Scribner's Sons, 1930. xi + 144 pp., 16 pls.
44. Miller, A. H. "Biotic Associations and Life-Zones in Relation to the Pleistocene Birds of California," *The Condor,* XXXIX (1938), 248–252. *See also* Grinnell (21).
45*a.* Moore, R. C. *Historical Geology.* New York, McGraw-Hill Book Company, Inc., 1933. xiii + 673 pp., 413 figs., 2 pls.
45*b.* Moore, R. C. *Introduction to Historical Geology.* New York, McGraw-Hill Book Company, Inc., 1949. ix + 582 pp., 364 figs., 21 pls. (A less detailed account than the previous work.)
46. Noble, L. F. "Structural Features of the Virgin Spring Area, Death Valley, California," *Bulletin of the Geological Society of America,* LII (1941), 941–999, 6 figs., 18 pls.
46*a.* Orowan, E. "The origin of the oceanic ridges." Scientific American, Nov., 1969. (see also September, 1969)
47. Peabody, F. E. "Reptile and Amphibian Trackways from the Lower Triassic Moenkopi Formation of Arizona and Utah," *University of California Publications, Bulletin of the Department of Geological Sciences,* XXVII (1948), 295–468, 40 figs.
48. Reed, R. D. *Geology of California.* Tulsa, American Association of Petroleum Geologists, 1933. xxiv + 355 pp., 60 figs., 1 pl.
49. Reed, R. D., and J. S. Hollister. *Structural Evolution of Southern California.* Tulsa, American Association of Petroleum Geologists, 1936. xix + 157 pp., 57 figs., 1 pl.
50. Richards, L. W., and G. L. Richards, Jr. *Geologic History at a Glance.* Stanford, California, Stanford University Press, 1934. 2 folded sheets.
51. Romer, A. S. *Vertebrate Paleontology.* Chicago, University of Chicago Press, 1966. ix + 687 pp., 377 figs.
52*a.* Sauer, C. O. "A Geographic Sketch of Early Man in America," *Geographic Review,* XXXIV (1944), 529–573, 3 figs.
52*b.* Sauer, C. O. "Environment and Culture During the Last Deglaciation," *Proceedings of the American Philosophical Society,* XCII (1948), 65–77, 1 fig.
52*c.* Savage, D. E. "Late Cenozoic Vertebrates of the San Francisco Bay Region," *University of California Publications, Bulletin of the Department of Geological Sciences,* XXVIII (1951), 215–314, 51 figs.
53. Schultz, C. B., and W. D. Frankforter. "The Geologic History of the Bison in the Great Plains," *Bulletin, University of Nebraska Museum,* III (1946), 1–10.
54. Sellards, E. H. "Early Man in America: Index to Localities and Selected Bibliography," *Bulletin of the Geological Society of America,* LI (1940), 373–432, 4 figs., 1 pl. [Continued in:] LVIII (1947), 955–978, 1 fig.
55. Sellards, E. H., G. L. Evans, and G. E. Meade. "Fossil Bison and Associated Artifacts from Plainview, Texas," *Bulletin of the Geological Society of America,* LVIII (1947), 927–954, 6 figs., 5 pls.
56. Shepard, F. P. *Submarine Geology.* New York, Harper & Brothers, 1948. xvi + 348 pp., 106 figs., 1 chart.

57. Shepard, F. P., and K. O. Emery. "Submarine Topography off the California Coast: Canyons and Tectonic Interpretation," *Geological Society of America, Special Papers,* No. 31 (1941). xiii + 171 pp., 42 figs., 18 pls., 4 charts.

58. Shimer, H. W. *An Introduction to the Study of Fossils.* New York, The Macmillan Company, 1933. xviii + 496 pp., 207 figs., 1 pl.

59. Simpson, G. G. *The Meaning of Evolution: A Study of the History of Life and of Its Significance for Man.* New Haven: Yale University Press, 1949. xv + 364 [+ i] pp., 38 figs.

60. Simpson, L. B. (trans.) *California in 1792: The Expedition of Jose Longinos Martinez.* San Marino, Huntington Library, 1938. xiii + 111 pp., map.

61. Stirton, R. A. "Cenozoic Mammal Remains from the San Francisco Bay Region," *University of California Publications, Bulletin of the Department of Geological Sciences,* XXIV (1939), 339–410, 95 figs.

62a. Stock, Chester. "Cenozoic Gravigrade Edentates [ground sloths] of Western North America," *Carnegie Institution of Washington, Publication* No. 331 (1925). xiii + 206 pp., 48 pls.

62b. Stock, Chester. "Rancho La Brea: A Record of Pleistocene Life in California," *Los Angeles Museum, Publication* No. 1 (1946). 82 pp., 27 figs.

62c. Stock, Chester. "Pushing Back the History of Land Mammals in Western North America," *Bulletin of the Geological Society of America,* LIX (1948), 327–332, 3 figs.

63a. Taliaferro, N. L. "Geologic History and Structure of the Central Coast Ranges of California," *State of California, Division of Mines, Bulletin* No. 118 (1941), 119–167, 5 figs., 1 pl.

63b. Taliaferro, N. L. "Franciscan-Knoxville Problem," *Bulletin of the American Association of Petroleum Geologists,* XXVII (1943), 109–219, 7 figs.

64. VanderHoof, V. L. "A Skull of Bison Latifrons from the Pleistocene of California," *University of California Publications, Bulletin of the Department of Geological Sciences,* XXVII (1942), 1–24, 5 figs., pls. 1–2.

65. Vokes, H. E. "How Old Is the Earth?" *American Museum of Natural History Guide Leaflet Series,* No. 75 (1941). 25 pp., 3 figs., and other illus.

66. Weaver, C. E., *et al.* "Correlation of the Marine Cenozoic Formations of Western North America," *Bulletin of the Geological Society of America,* LV (1944), 569–598, pl. 1.

67. Welles, S. P. "Elasmosaurid Plesiosaurs, with Description of New Material from California and Colorado," *Memoirs of the University of California,* XIII (1943), 125–254, 37 figs., frontis., pls. 12–29.

67a. Welles, S. P. "A new species of elasmosaur from the Aptian of Colombia and a review of the Cretaceous plesiosaurs." Univ. Calif. Publications in Geological Sciences, vol. 44, 1–96, illustr. 1962.

67b. Welles, S. P., and J. Cosgriff. "A revision of the labyrinthodont family Capitosauridae." Univ. Calif. Publications in Geological Sciences, vol. 54, 1–148, illustr.

67c. Welles, S. P., and Estes, R. *Hadrokkosaurus bradyi* from the upper Moenkopi Formation of Arizona. Univ. Calif. Publications in Geological Sciences, vol. 84, 1–56, illustr. 1969.

68a. Williams, Howel. "The History and Character of Volcanic Domes," *University of California Publications, Bulletin of the Department of Geological Sciences,* XXI (1932), 51–146, 37 figs.

68b. Williams, Howel. "Geology of the Lassen Volcanic National Park, California," *ibid.,* XXI (1932), 195–385, 64 figs., 3 maps.

68c. Williams, Howel. "Calderas and Their Origin," *ibid.,* XXV (1941), 239–346, 37 figs.

68d. Williams, Howel. "Volcanoes of the Three Sisters Region, Oregon Cascades," *ibid.,* XXVII (1944), 37–84, 4 figs., pls. 4–12, map.

68e. Williams, Howel. *The Ancient Volcanoes of Oregon.* Eugene, Oregon, State System of Higher Education, 1948. x + 55 pp., 9 figs., 13 pls.

69. Wood, H. E., *et al.* "Nomenclature and Correlation of the North American Continental Tertiary," *Bulletin of the Geological Society of America,* LII (1941), 1–48, 1 pl.

70. Wormington, H. M. Ancient Man in North America. *Colorado Museum of Natural History, Popular Series,* No. 4 (1944). 89 pp., illus. (2d ed., 1949, 198 pp.)

71. Zeuner, F. E. *Dating the Past: An Introduction to Geochronology.* London, Methuen, 1946. xviii + 444 pp., 103 figs., 24 pls.

188

Index

Age of the earth, 28
Ages of life, 15
Agriculture, New World, 137
Algae, 29, 39, 40
Algonkian period, 29, 165, 167
Ammonites and ammonoids, 47, 59,
    62, 63, 74, 75
Amphibia, 47, 49, 54, 59, 66
Antelopes, 92, 93, 98, 109
Aplodontia, 92
Archaeopteryx, 74
Archaen period, 29, 165, 167
Argus Range, 52, 58
Arizona, 46
Armadillos, 86, 99
Artifacts, 110, 133, 138, 141, 176
Artiodactyls, 84
Asphalt, 18, 113, 115, 140
Atlatl dart, 136, 137
Australian mammals, 85
Australopithecus, 124

Bacteria, 29, 38
Barnacles, 86
Basin ranges, 103
Basketmakers, 110, 137
Bear dogs, 95, 98
Bear Flag revolt, 144
Bears, 86, 99, 106, 107
Beavers, 93, 98, 106
Bennison, Allan, 18
Berkeley Hills, 25, 98, 153
Bighorn, 106, 110
Birds, 16, 49, 74, 75, 84, 115, 116,
    118, 119
Bison, 17, 18, 19, 106, 108, 118, 127,
    133, 134
Black Hawk quarry, 98
Blastoids, 42, 47
Borax Lake, 176
Bow and arrow, 137, 139
Brachiopods, 40, 42, 47, 52
Brokeoff Mountain, 154
Bryozoans, 42

Cabrillo, Juan Rodriguez, 143
Cajon Pass, 19
Calamites, 53
Calaveras skull, 137
Calistoga petrified forest, 98
Cambrian, 19, 33, 34, 39, 41, 165, 167
Camels, 18, 92, 93, 98, 99, 106, 110, 138
Canada porcupine, 86
Carbon-14 dating, 139, 176
Carboniferous, 52–55
Carnivores, 59, 75, 85, 158
Carp, 74
Carpinteria tar pits, 119, 174
Cats, 85, 87, 93, 98, 99, 106, 118
Cermeño, Sebastian Rodriguez, 144
Channel Islands, 80, 107, 140, 141
Chirotherium, 66
Chronology, 28, 164, 174, 176
Chumash Indians, 140, 141
Chumasius, 87
Clams, 42, 47
Clark, Bruce L., 98
Climactic changes, 47, 49, 80, 81, 87,
    109, 110–11, 159, 176
Cloud-condensation theory, 23
Clovis artifacts, 176
Club mosses, 47, 53
Coal beds, 54, 87
Coal forests, 53
Coast Ranges, 25, 26, 27, 33, 69, 72
Cochise culture, 176
Cockroaches, 54, 59
Colorado Desert, 86
Conifers, 58, 67, 72, 119, 158
Continental deposits, 169
Corals, 40, 41, 43, 45, 47, 52, 63, 74
Cordilleran geosyncline, 33, 41, 44
Cotylosaurs, 59
Coyotes, 99, 106
Crabtree, Don, 134
Crater Lake, 155, 176
Creodonts, 87
Cretaceous, 26, 33, 54, 72–73, 74–81,
    164–65, 166, 171

Crinoids, 40, 47, 52, 74
Crocodiles, 72, 75
Cro-Magnons, 123–24
Crossopts, 47, 74
Crustal movements, 25, 27, 32, 35,
    103, 149, 152
Cycads, 67, 69, 76, 92
Cystoids, 42, 47

Darwin Hills, 52, 58
Death Valley, 96, 103
Deer, 84, 93, 98, 106
Devonian, 46–47, 49, 53, 165, 167
Diatomite, 81, 86
Diatoms, 81, 86
Diatryma, 87
Dicynodont reptiles, 70–71
Dinosaurs, 18, 19, 72–73, 75, 79, 80
Disease carriers, 156–57
Dogs, 87, 98, 106
Dragonflies, 54
Drake, Francis, 144
Dwarfing of island animals, 109

Earth:
    age of, 28
    history of, 19, 27, 33, 35
    origin of, 23
Earthquakes, 25, 149, 152
Edentates, 84
Elephants, 17, 86, 105, 107, 109
Energy:
    atomic, 23
    chemical, 38
    radiant, 28
    of universe, 32
Eocene, 54, 84, 87, 165–71
Eohippus, 87
Erosion, 33, 73
Eurypterids, 42
Evolution, 16, 28, 39, 49, 85, 123–25, 172
Extermination, 49, 80, 109, 156–57
Extinction, 28, 47, 75, 99, 107, 111

Farallones, 25
Fault zones, 25, 103, 149–53
Ferns, 46, 47, 53, 67, 69, 76
Fishes, 16, 45, 46, 47, 48, 49, 74
Flowering plants, 75, 81
Folsom:
    man, 126–34

points, 134–35
sites, 134, 174
Footprints, 66
Faraminifers, 52, 53, 59, 81, 86
Forest fires, 157
Fossils:
    faunas and floras, 61, 92–93, 109,
        168–69
    geographic distribution, 61
    horizons, 168
    records, 15, 19, 27, 33, 35, 37,
        39, 72, 85, 87, 98
Foxes, 98
Franciscan:
    gulf, 33, 69, 80
    series, 26, 69, 72–73, 80, 88
Fresno, 80
Frogs, 47, 98

Ganoids, 47, 74
Geologic:
    wents, 72–73, 86, 164–71
    time scale, 164–65
    records, 35
Geosyncline, 24, 32, 41, 44, 72
Glacial stages, 168–69
Glyptodonts, 86, 99
Goldfish, 74
Gold rush, 145
Gomphotherium, 98
Grand Canyon, 29, 39, 41
Granitic basement, 24, 26, 152–53
Graptolites, 41, 44, 47
Grasses, 92
Grazing ranges, 159
Ground sloths, 18, 86, 93, 99, 109, 116
Ground squirrels, 98
Guide fossils, 39
Gypsum Cave, 107, 109–10, 176

Heizer, Robert F., 139
History of life, 26, 27–29, 35, 39,
    52–54, 173
Hohokam culture, 137, 176
Horses, 18, 87, 92, 93, 97, 98, 99, 106
Horsetails, 46, 53, 67, 72, 76
Human evolution, 121–25
Hydrozoans, 41, 47, 81

Ice stages, 102–3
Ichthyosaurs, 62–63, 69, 74, 75
Ictidosaurs, 68

Immigration to California, 138, 144–45
Indians, 137–42
Insects, 49, 54, 59, 74, 81
Inyo Mountains, 29, 41, 46, 52, 58
Irvington fauna, 107, 109
Isostasy, 32

Jaguars, 18, 118
Java man, 124
Jurassic, 33, 69, 72–73, 76–77, 166, 171

Kaibab limestone, 58
Klamath Mountains, 33, 46, 69
Knoxville flora, 72

Laramide revolution, 68, 75
Lassen Peak, 25, 154, 155
Leonard Rockshelter, 176
Life, origin of, 31
Lizards, 16, 68, 72, 75, 109
Longinos Martínez. See Martínez.
Los Angeles fossils, 18, 138, 176
Lungfishes, 47, 49, 68
Lycopods, 53

McKittrick, 119
Mammals, 17, 18, 68, 75, 80, 83–87,
        92–99, 105–11
Mammoths, 17, 107, 116, 118
Man, 85, 121–42, 138, 139, 146
Man ape, 74, 124
Mankato glacial stage, 176
Marble Mountains, 41, 58
Marine:
        faunas, 61–63
        fossils, 27, 35, 86–87
        formations, 72–73, 170, 171
        terraces, 27
Mariposa slates, 69
Martínez, José Longinos, 18
Mason, Herbert L., 158
Mastodonts, 18, 93, 98, 106, 107
Mexican civilizations, 138
Migration of floras, 109
Miocene, 86, 90–91, 92, 97, 107, 158,
        165, 166, 169, 171
Miohippus, 87
Mississippian, 33, 52
Moenkopi formation, 66
Mojave Desert, 25, 29, 52, 138
Mososaurs, 18, 78, 81

Mother Lode, 69, 145
Mountain formation, 26–27, 68, 75
Mountain goats, 107
Mount Diablo, 17, 87, 98
Mount Mazama, 155

Natural resources, 159
Nautilus and nautiloids, 42, 47, 60, 62, 74
Navajo formation, 73
Neanderthals, 124
Nopah hills, 41, 46

Oil deposits, 148–49
Olenellus, 41, 43
Oligocene, 87, 96–97, 165, 166, 171
Ordovician, 41, 42, 45, 165, 167
Oregon caves, 176
Ostracoderms, 42, 45, 47
Owens Valley, 26, 103, 153
Oxen (Euceratherium), 107, 119
Oysters, 86, 87

Painted Desert beds, 66, 67, 68
Paleocene, 75, 84, 85, 165, 166, 169, 171
Paleozoic, 33, 34, 39, 41, 165, 167
Panamint Mountains, 43, 46, 103
Peccaries, 93, 98
Peking man, 124
Pelycosaurs, 59
Pennsylvanian, 33, 52, 165, 167
Perissodactyls, 84
Permian, 33, 58, 59, 165, 167
Petrified forests, 67, 98
Petroleum deposits, 25, 115, 119, 148–49
Philippine galleons, 143
Phytosaurs, 68
Pig deer (oreodonts), 87, 98
Pithecanthropus, 124
Placerias, 68, 70–71
Plants, 39–40, 46–47, 49, 87, 138, 159
Pleistocene:
        chronology, 164–65, 175
        epoch, 102–3
        faunas and floras, 104ff.
Plesiosaurs, 18, 75, 78, 81
Pliocene, 86, 98–99, 107, 148, 165,
        166, 169, 171
Pliohippus, 93
Population problems, 145
Porcupines, 84, 86
Poway conglomerates, 87

Primates, 85, 123
Pronghorn antelopes, 106, 118
Protoplasm, 38
Protozoa, 40, 81
Providence Mountains, 41, 58
Pterodactyls, 74
Pueblo Culture, 110
Pyramid Lake, 29

Rabbits, 98
Raccoons, 98
Radiolarians, 29, 41, 69
Rancho La Brea, 18, 107, 115–19
Recent chronology, 176
Redwoods, 158
Reptiles, 56–81
Rhinoceros, 87, 98
Ring-tailed cats, 98
Rodents, 84, 87
Rodriguez Cabrillo. See Cabrillo.
Rodriguez Cermeño. See Cermeño.
Russians in California, 144

Saber cats, 18, 85, 93, 95, 98, 115,
    116, 118
Salamanders, 16, 47, 49, 54
San Andreas rift, 149–53
San Bernardino Mountains, 26, 69, 103
Sandia culture, 135
San Dieguito site, 139
San Gabriel Mountains, 26
San Jacinto Mountains, 26
San Joaquin Valley, 27, 33, 79, 86,
    89, 98, 138
San Mateo flora, 107
Santa Barbara harbor, 147
Scorpions, 43, 49
Sea salts and metals, 146
Sea urchins, 42, 47
Seals, sea lions, sea cows, 93
Seaways, ancient, 33, 60–62, 69,
    86, 88–91
Sediments, 24, 25, 26, 32–35, 58, 69, 72
Seed ferns, 53
Sequence of life forms, 39, 164–65
Sequoia, 16
Serra, Junipero, 144
Sespe beds, 87
Shellfish, 74, 86–87
Shell mounds, 141
Sierra Nevada, 25, 26, 33, 69, 72–73, 102, 103

Silurian, 43–45, 165, 167
Snails, 40, 42, 49
Social evolution, 125
South African man apes, 124
South American faunas, 85, 106
Spanish exploration, 143
Sponges, 40, 41, 42, 47
Squirrels, 98
Stone lilies (crinoids), 40, 42, 74
Submarine canyons, 148, 151

Tamiosoma, 86
Tapirs, 18, 106
Tar pits, 18, 113, 115, 119, 174
Taylorsville, 43, 46
Tecopa region, 29
Telescope Range, 41
Teratornis, 116, 118
Termites, 81
Tertiary, 33, 86, 87, 157, 165, 166, 169
Thalattosaurs, 63
Toothed birds, 81
Topango site, 139
Toroweap limestone, 58
Tortoises, 93, 109
Tranquility site, 138
Tremarctotherium, 107
Triassic, 33, 60–68, 165, 167
Trilobites, 19, 40–41, 42, 43, 59
Treplopus, 87
Tufa domes, 29, 103
Tyrannosaurus, 75

Uranium clock, 28

Ventana Cave, 109, 176
Viruses, 38
Vizcaino, Sebastian, 144
Volcanoes, 26, 98, 153–55

Water shortage, 146
Weasels, 98, 106
Williams, Howel, 155
Wilmington harbor, 155
Windmiller mound, 139
Wingate formation, 73
Wisconsin glaciation, 176
Wolverines, 106
Wolves, 18, 92, 110, 115, 119, 127

Yosemite Valley, 102